THE FINITE ELEMENT METHOD
IN MECHANICAL DESIGN

Charles E. Knight, Jr.
*Virginia Polytechnic Institute
and State University*

PWS-KENT Publishing Company
Boston

PWS–KENT
Publishing Company

20 Park Plaza
Boston, Massachusetts 02116

PWS-KENT Publishing Company is a division of Wadsworth, Inc.

Library of Congress Cataloging-in-Publication Data

Knight, Charles E.
  The finite element in mechanical design / Charles E. Knight, Jr.
  p.   cm.
  Includes bibliographical references and index.
  ISBN   0-534-93187-1
  1. Finite element method. 2. Engineering design--data processing.
  3. Computer-aided design   I. Title
TA347.F5K65   1993   92-10374
620'.0042'0151535 -- dc20   CIP

 This book was printed on recycled, acid-free paper.

Printed in the United States of America

  94 95 96 97 - 10 9 8 7 6 5 4 3

Sponsoring Editor: Jonathan Plant
Assistant Editor: Mary Thomas
Production Editor: Kirby Lozyniak
Cover Design: Monique Calello
Manufacturing Coordinator: Ellen Glisker
Text Printer/Binder: Arcata Graphics/Halliday
Cover Printer: Henry Sawyer Company

PREFACE

The *Finite Element Method in Mechanical Design* was written to support the finite element program user. Emphasis is on the practical aspects of proper modeling, checking, and interpretation of results. Theoretical aspects are introduced as they are needed to help understand the operation. The text covers most of the typical engineering structures from the simple to the more complex.

The Finite Element Method in Mechanical Design results from notes developed while teaching an undergraduate elective course in finite element practice for the past eight years. Through this period there have been dramatic improvements in both hardware and software computer capability. Mostly, commercial software of various brands have been used in order to conduct realistic analyses.

The finite element method has rapidly become a vital tool for analysis of mechanical designs. It has reached the point where practically every design engineer has access to a finite element program through their company's mainframe computer or on a micro-computer. This method is extremely powerful in terms of the many different types of problems it can solve. The method applies to many engineering fields, however this book concentrates on the application to mechanical design.

Use of this tool does not guarantee correct results. It is a numerical procedure involving approximations of theoretical behavior. In order to produce correct results the model must be designed correctly, and be able to reach numerical convergence (assuming the computer program is without error). Correct results then depend primarily on the user's ability to utilize the tool. It in no way supplants the engineer's responsibility to do approximate engineering calculations, use good design practice, and apply engineering judgment to the problem. Instead, it should supplement these skills to ensure that the best design is obtained.

With the ready access of finite element analysis comes the need to provide the understanding required to accurately and effectively use the method. Most design engineers are not going to have the time nor the inclination to study all the theoretical formulations and computer algorithms much less write their own program. Therefore, most engineers take the role of users of an in-house or commercial program.

Organization of This Text

Chapters 1 through 9 present methods in static analysis of various classes of structures. Chapter 1 gives a simple introduction of the finite element method. Chapter 2 shows the fundamental relationships between classes of engineering structures and corresponding finite elements. Chapters 3 through 9 apply the finite element method to each class of structure.

In each of these chapters, the finite element analysis of the given structure class is discussed. First, formulations of the elements that fit the given structure are developed. This is important for understanding the nature and degree of approximation in the resulting solution. A discussion of the steps taken to perform an analysis from model conceptualization to evaluation of results then follows. In this way the student can get through a complete analysis very early in the study, and should be running in conjunction some software package to do analysis projects throughout this study. Obviously some of the steps are common for every class of structure. Unique features pertinent to each class are introduced in consecutive chapters. Each chapter also includes case studies in which a model is constructed, an analysis is run, and the results for one or more examples is presented and evaluated.

Chapter 10 introduces dynamic analysis. Included in this chapter are descriptions of the types of dynamic analyses and the element mass matrix formulations. Also presented are the differences from static modeling techniques. Chapter 11 brings in some special modeling techniques, and Chapter 12 presents heat transfer by the finite element method and the analysis of thermal stresses.

Information About Accompanying Disk

A PC computer disk is provided with this book which contains a finite element program called FEPC. The program runs on an IBM PC or compatible computer. It will solve two-dimensional problems using truss, beam, plane stress, plane strain or axisymmetric solid elements. Dimensional limits of the program are described in an abbreviated user's guide in the Appendix. Also, you should review the README file on the disk for startup instructions. The complete user's guide is in a file on the disk. The program is provided as shareware.

Although technical support can not be provided, we will replace any defective disk. Also, you may contact the author regarding any program errors or availability of program updates.

Acknowledgments

I would like to thank my former students in the course for their suggestions in refining the material for this text. Thanks to my colleagues and department head in the Department of Mechanical Engineering for their encouragement and support. Professor Doug Nelson reviewed the heat transfer chapter and gave me several improvements. Matthew Wicks did outstanding work on much of the graphics in the text and in the solutions manual.

The staff at PWS-KENT Publishing Company have been exceptionally helpful and very easy to work with. I must thank my editor, Jonathan Plant, for his enthusiasm, encouragement, and patience throughout this project. Patty Adams provided very clear layout and startup guidelines for the text. Kirby Lozyniak made the text much smoother and better arranged. Mary Thomas provided much needed organizational help.

Finally, thanks to the following individuals who reviewed this text and provided many useful comments and suggestions that helped to improve the final product:

Eduardo Bayo
University of California-Santa Barbara

Camille A. Issa
Mississippi State University

Fernando E. Fagundo
University of Florida

Stephen R. Swanson
University of Utah

Charles E. Knight, Jr.

This book is dedicated to my wife, Ann

CONTENTS

CHAPTER 4

BEAMS AND FRAMES 65

CHAPTER 5

TWO-DIMENSIONAL SOLIDS 86

CHAPTER 6

THREE-DIMENSIONAL SOLIDS 134

C H A P T E R 1

THE FINITE ELEMENT METHOD

1.1 General Overview

The finite element method is enjoying widespread use in many engineering applications. Although first developed for structural analysis, it now solves problems in heat transfer, fluid mechanics, acoustics, electromagnetics, and other specialized disciplines. In conduction heat transfer, we solve for the temperature distribution throughout the body with known boundary conditions and material properties whether steady state or time dependent. Application to fluid mechanics begins with steady inviscid incompressible flow and progresses to very complex viscous compressible flow. The whole area of computational fluid dynamics has made rapid progress in recent years. Acoustics is another area where great strides are being made based on finite element and boundary element numerical methods. Electromagnetic solutions for magnetic field strength provide insight for design of electromagnetic devices. Many of these capabilities are now being coupled to yield solutions to fluid-structure interactions, convective heat transfer and other coupled problems.

The finite element method is a numerical method for solving a system of governing equations over the domain of a continuous physical system. The method applies to many fields of science and engineering, but this text focuses on its application to structural analysis. The field of continuum mechanics and theory of elasticity provide the governing equations.

The basis of the finite element method for analysis of solid structures is summarized in the following steps. Small parts called *elements* subdivide the domain of the solid structure illustrated in Figure 1-1. These elements assemble through interconnection at a finite number of points on each element called *nodes*. This assembly provides a model of the solid

1

structure. Within the domain of each element we assume a simple general solution to the governing equations. The specific solution for each element becomes a function of unknown solution values at the nodes. Application of the general solution form to all the elements results in a finite set of algebraic equations to be solved for the unknown nodal values. By subdividing a structure in this manner, one can formulate equations for each separate finite element which are then combined to obtain the solution of the whole physical system. If the structure response is linear elastic, the algebraic equations are linear and are solved with common numerical procedures.

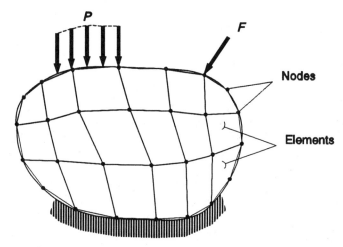

Figure 1-1. Two-Dimensional Continuum Domain

Since the continuum domain is divided into finite elements with nodal values as solution unknowns, the structure loads and displacement boundary conditions must translate to nodal quantities. Single forces like F apply to nodes directly while distributed loads like P are converted to equivalent nodal values. Supports like the grounding indicated by the hatch in Figure 1-1 resolve into specified displacements for the supported nodes.

At least two sources of error are now apparent. The assumed solution within the element is rarely the exact solution. The error is the difference between assumed and exact solutions. The magnitude of this error depends on the size of the elements in the subdivision relative to the solution variation. Fortunately, most element formulations converge to the correct solution as the element size reduces. The second error source is the precision of the algebraic equation solution. This is a function of the computer accuracy, the computer algorithm, the number of equations, and the element subdivision.   Both error sources are reduced with good modeling practices.

In theory all solid structures could be modeled with three-dimensional solid continuum elements.   However, this is impractical since many structures are simplified with correct assumptions without any loss of accuracy, and to do so greatly reduces the effort required to reach a solution.   Different types of elements are formulated to address each class of structure.   Elements are broadly grouped into two categories, structural elements and continuum elements.

Structural elements are trusses, beams, plates, and shells.   Their formulation uses the same general assumptions about behavior as in their respective structural theories.   Finite element solutions using structural elements are then no more accurate than a valid solution using conventional beam or plate theory for example.   However, it is usually far easier to get a finite element solution for a beam, plate or shell problem than it is using conventional theory.

Continuum elements are the two- and three-dimensional solid elements. Their formulation basis comes from the theory of elasticity.   The theory of elasticity provides the governing equations for the deformation and stress response of a linear elastic continuum subjected to external loads.   Few closed form or numerical solutions exist for two-dimensional continuum problems, and almost none exist for three-dimensional problems which makes the finite element method invaluable.

An extensive literature has developed since the 1960s when the term "finite element" originated.   The first textbook appeared in 1967[See Reference 1.1].   The number of books and conference proceedings published since then is near two hundred and the number of journal papers and other publications is in the thousands.   The engineer beginning study of the finite element method may consult references [1.2], [1.3], [1.4], [1.5], [1.6], [1.7], or [1.8] for formulation fundamentals, [1.9], [1.10], [1.11], [1.12], [1.13], or [1.14] for structural and solid mechanics applications, and [1.15], [1.16], or [1.17] for computer algorithms and implementation.

1.2 One-Dimensional Spring System

The fundamental operation of the finite element method is illustrated by analysis of a one-dimensional spring system.   A two-spring structure is sketched in Figure 1-2.   Each spring is an element identified by the number in the box.   The spring elements have a node at each end and they connect at a common node.   The number in the circle labels each node.   The number in the box labels each element.   The subscripted u values are the node displacements, i.e. degrees-of-freedom.   There is an applied force F

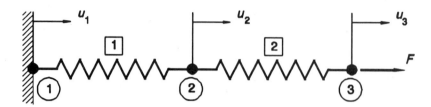

Figure 1-2. One-Dimensional Spring Structure

at node 3. We wish to solve for the node displacements and spring forces.

The first step is to formulate a general element. Figure 1-3 shows a spring element. The element label is p with nodes i and j. For a displacement formulation assume positive displacement components of u_i at node i and u_j at node j. The element has a spring constant k, so node forces result when these displacements occur. Define f_{ip} as the force acting on node i due to the node displacements of element p. Application of simple equilibrium forms equations (1.1).

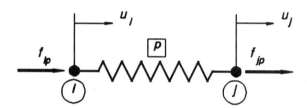

Figure 1-3. One-Dimensional Spring Element

$$f_{ip} = -k_p u_i + k_p u_j$$
$$f_{jp} = k_p u_i - k_p u_j \tag{1.1}$$

We write these in matrix form in equation (1.2) and then symbolically in equation (1.3).

$$\begin{bmatrix} k_p & -k_p \\ -k_p & k_p \end{bmatrix} \begin{Bmatrix} u_i \\ u_j \end{Bmatrix} = \begin{Bmatrix} -f_{ip} \\ -f_{jp} \end{Bmatrix} \tag{1.2}$$

$$[k]\{d\} = \{f\} \tag{1.3}$$

Here, $[k]$ is the element stiffness matrix, $\{d\}$ is the element node displacement vector, and $\{f\}$ is the element node internal force vector. These steps complete the element formulation.

Now apply the general formulation to each element:

for element 1

$$\begin{bmatrix} k_1 & -k_1 \\ -k_1 & k_1 \end{bmatrix} \begin{Bmatrix} u_1 \\ u_2 \end{Bmatrix} = \begin{Bmatrix} -f_{11} \\ -f_{21} \end{Bmatrix} \tag{1.4}$$

and, for element 2

$$\begin{bmatrix} k_2 & -k_2 \\ -k_2 & k_2 \end{bmatrix} \begin{Bmatrix} u_2 \\ u_3 \end{Bmatrix} = \begin{Bmatrix} -f_{22} \\ -f_{32} \end{Bmatrix}. \tag{1.5}$$

The force components in the element equations are internal forces on the nodes produced by the elements when the nodes displace. Equilibrium requires that the sum of the internal forces equals the external force at each node. Representing the external force by F_i, where i represents each node, the equilibrium equations become:

$$
\begin{aligned}
\text{at node 1} \quad & \sum \text{forces} = 0 \quad \Rightarrow \quad -f_{11} = F_1 \\
\text{at node 2} \quad & \sum \text{forces} = 0 \quad \Rightarrow \quad -f_{21} - f_{22} = F_2 \\
\text{at node 3} \quad & \sum \text{forces} = 0 \quad \Rightarrow \quad -f_{32} = F_3
\end{aligned}
\tag{1.6}
$$

Substitute the element equations for the internal force terms in the equilibrium equations (1.6), and that, in effect, performs the structure assembly and yields the structure equations (1.7).

These are written in matrix form in equation (1.8) and symbolically in equation (1.9) where, $[K]$ is the structure stiffness matrix, $\{D\}$ is the structure node displacement vector, and $\{F\}$ is the structure external force vector.

The set of structure or system equations must now be solved. The spring constants of the springs are known so all terms in the structure

$$k_1 u_1 - k_1 u_2 \qquad\qquad = F_1$$
$$-k_1 u_1 + k_1 u_2 + k_2 u_2 - k_2 u_3 = F_2 \qquad (1.7)$$
$$- k_2 u_2 + k_2 u_3 = F_3$$

$$\begin{bmatrix} k_1 & -k_1 & 0 \\ -k_1 & k_1+k_2 & -k_2 \\ 0 & -k_2 & k_2 \end{bmatrix} \begin{Bmatrix} u_1 \\ u_2 \\ u_3 \end{Bmatrix} = \begin{Bmatrix} F_1 \\ F_2 \\ F_3 \end{Bmatrix} \qquad (1.8)$$

$$[K]\{D\} = \{F\} \qquad (1.9)$$

stiffness matrix are known. The applied forces are known and the node displacements become the unknowns in this set of three simultaneous equations. We get the solution by premultiplying both sides of the equation (1.9) by the inverse of $[K]$. However, in this case the inverse of $[K]$ is singular, meaning that we cannot get a unique solution. Physically this means that the structure can be in equilibrium at any location in the x space, and it is free to occupy any of those positions. This allows rigid body motion. To have a unique solution we must locate the structure, that is, apply boundary conditions such as a fixed displacement on one of the nodes which is enough to prevent rigid body motion.

If an external force F applies to node 3, and the spring attaches to the wall at node 1, then it is natural to set the displacement of node 1 to zero. This action zeroes the first column of terms in the structure stiffness matrix, and that leaves three equations with two unknowns. If the value of the reaction force at node 1 is unknown then we may skip the first equation and choose the second and third equations to solve for the unknown displacements. If the external force on node 2 is zero then

$$\begin{bmatrix} k_1+k_2 & -k_2 \\ -k_2 & k_2 \end{bmatrix} \begin{Bmatrix} u_2 \\ u_3 \end{Bmatrix} = \begin{Bmatrix} 0 \\ F \end{Bmatrix} \qquad (1.10)$$

Now we may get the solution of the resulting equations (1.10) by premultiplying both sides of the equation by the inverse of this reduced structure stiffness matrix. Using the solved displacements, we calculate each element internal force by use of the individual element equations. In this example the force calculation is trivial, but in more complex elements this step determines stresses in elements of the structure.

Also, in this example the calculation of the reaction force at node 1 is easily done from elementary equilibrium equations. However in general cases, the structure under analysis may be statically indeterminant and the reaction forces at locations of support or fixed displacement may not be known. In that situation, the equations involving these reaction forces are stored and solved after the displacements are found to calculate any desired reaction forces.

This detailed example illustrates most of the fundamental steps in the finite element method. The finite element method obviously overpowers the example case. However, a complex spring arrangement could use the procedure for analysis, or if a computer program was written, solution for a complex spring arrangement would come quickly with input of the spring constants and connectivity. The major differences between this example and actual practice are that, (1) nodes usually have more than one displacement component or degree-of-freedom, (2) the element formulation is chosen to match the class of structure being analyzed, and (3) there are a large number of equations to be solved.

1.3 Using a Computer Program

There are three stages which describe the use of any existing finite element program. The preprocessing stage creates the model of the structure from inputs provided by the analyst. A preprocessor then assembles the data into a format suitable for execution by the processor in the next stage. The processor is the computer code that generates and solves the system equations. The third stage is postprocessing. The solution in numeric form is very difficult to evaluate except in the most simple cases. The postprocessor accepts the numeric solution, presents selected data, and produces graphic displays of the data that are easier to understand and evaluate.

Figure 1-4 draws a block diagram of a typical finite element computer program. Before entering the program's preprocessor, the user should have planned the model and gathered necessary data. In the pre-processor block, the user defines the model through the commands available in the preprocessor. The definition includes input and generation of all node point coordinates, selection of the proper element from the program's element library, input and generation of node connectivity to define all elements, input of material properties, and specifying all displacement boundary conditions, loads and load cases. The completion of the preprocessing stage results in creation of an input data file for the analysis processor.

The processor reads from the input data file each element definition, calculates terms of the element stiffness matrix, and stores them in a data

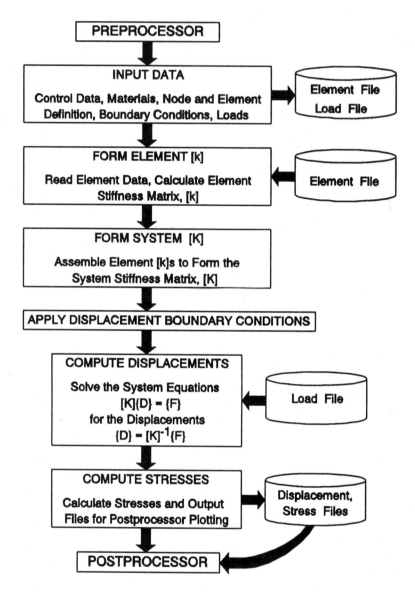

Figure 1-4. Finite Element Computer Program Block Diagram

array or on a disk file. The element type selection determines the form of the element stiffness matrix. The next step is to assemble the structure stiffness matrix by matrix addition of all element stiffness matrices. The application of enough displacement boundary conditions to prevent rigid body motion reduces the structure stiffness matrix to a nonsingular form.

Then the equation solution may be done with several different computer algorithms, but most use some variation of Gaussian elimination. For a large set of equations this is the most computationally intensive step in the process. The solution here yields values for all node point displacement components in the model. The node displacements associated with each element combined with the element formulation matrix yield the element strains. The element strains with the material properties yield the stresses in each element. The processor then produces an output listing file with data files for postprocessing.

Postprocessing takes the results files and allows the user to create graphic displays of the structural deformation and stress components. The node displacements are usually very small for most engineering structures so they are magnified to show an exaggerated shape. Node displacements are single-valued, but node values of stress are multi-valued if more than one element is attached to a given node. Node stress values are usually reached by extrapolation from internal element values and then averaged for all elements attached to the node. Contour plots or other stress plots desired by the user are created from the node values. In some post-processing programs criterion plots of the factor of safety, stress ratio to yield, or stress ratio to allowable values are also generated and displayed.

The engineer is then responsible for interpreting the results and taking whatever action is proper. The user must estimate the validity of the results first. This is very important because the tendency is to accept the results without question. Experience, thorough checking of the modeling assumptions and resulting predicted behavior, and correlation with other engineering calculations or experimental results all contribute to estimating the validity of the results.

1.4 The Analysis Step-by-Step

An idealized analysis procedure will bring all steps into proper order. The procedure includes initial planning, deciding if a finite element analysis is needed, doing a needed analysis, and presenting the results. Following this procedure should allow the user to perform an effective and efficient analysis. A block diagram is shown in Figure 1-5 to organize the following narrative.

When the engineer has created a conceptual design ready for analysis or has an existing design failure needing analysis, the first step is to define the analysis problem as clearly as possible. The definition includes the determination of what type of analysis is to be performed (static, dynamic, etc.), whether the solution will be two- or three-dimensional, and what the criteria will be for the analysis.

The criteria for the analysis identify the important variables for evaluation. These could include the maximum stress, the average stress,

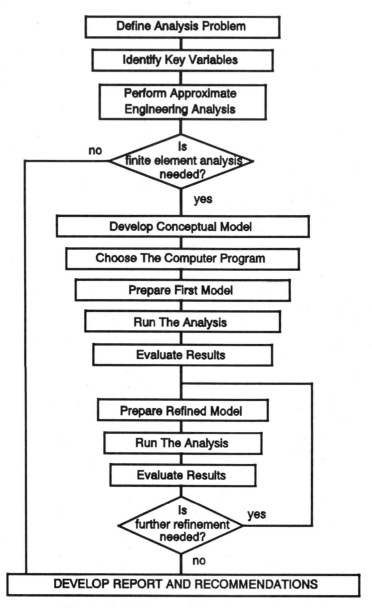

Figure 1-5. The Analysis Step-by-Step

the strain, the deformation, fracture load, yield load, critical stress location, or several other factors. The most critical of these serve to guide the modeling and to guide the presentation of results. For example, if a particular location in a new design is expected to be critical or has been critical in an existing design, tailor the model to examine that location carefully.

The time allowed to complete the analysis may determine whether an adequate analysis can be done at all. If time constraints are too tight, then you must sacrifice accuracy, and the analysis would thus only show the qualitative aspects of the design. The required solution accuracy established at the beginning helps to estimate the number of analysis cycles. These parameters influence modeling decisions and determine the overall scope of the analysis.

After completion of the problem definition, an approximate engineering analysis should follow. Determine the design load conditions and expected, or possible, overload conditions. Determining these load conditions should include some estimate of the accuracy of the loading. Find the material property data including some estimate of its statistical variation. If the variation is high, some sensitivity cases may have to be run to check the material property influence on the most critical analysis results. Approximate the structure by similar structures that have known solutions, such as beams, thick cylinders, simple plates, or others. Using these approximate solutions, estimate the values of the critical variables identified in the problem definition step above. Based on these approximate results and the degree of confidence placed in the approximate solutions, determine if you need to do a finite element analysis.

If the decision is yes, begin by developing a conceptual finite element model. The analyst developing conceptual model will lay out the geometry of the model, choose the kind of element that would be best suited, and roughly plan the mesh for the model. Examine the structure for any symmetries that exist in the geometry and loading conditions, and from those symmetries select a repeating section that will define the model geometry. Exploiting symmetry reduces the effort required to create a model as well as reducing the number of computations and increasing the numerical accuracy.

Determine if the solution is to be two-dimensional or three-dimensional. If the problem requires a three-dimensional model, first make a two-dimensional simplification for the initial analyses. A two-dimensional analysis should always preceed a three-dimensional analysis to gain insight into the nature of the problem and provide some experience for developing a good three-dimensional model.

For the section model choose the type of finite element formulation needed to represent correctly the structural behavior. Also, identify any supporting element types needed for special interfaces or boundary conditions. Finally, sketch a rough mesh plan making an effort to lay out the element subdivisions according to the expected solution variations.

Completion of the conceptual model will provide the information needed to choose the proper computer program to use. Availability may dictate the choice of program to use, however if more than one program is available then the following factors should enter into the choice. Of most importance is that the type of element needed to correctly model the structure be available in the program's element library with any special element types called for in the conceptual model. If any significant material or geometric nonlinearities potentially exist in the solution, choose a program that can produce the nonlinear solution. However, a linear solution should always preceed any try at a nonlinear solution.

The next features to consider in choosing the program involve the mesh generation and preprocessing abilites and the postprocessing and graphic displays available for presenting the results. Since time is usually very important and readily understandable results are valuable, these features do a great deal toward aiding the engineer in processing the information.

After making the program choice, it is time to prepare the first model. It begins by developing a detailed mesh plan that includes the degree of refinement desired in the mesh at all critical locations. The analyst gathers and assembles all the required data and input information for the preprocessor. At this point, the overall geometry of the model section is input with specification of the mesh generation. The actual mesh generation follows, and if the mesh is considered acceptable, the boundary conditions for the enforced displacements are applied. Also, within the preprocessor the load case or cases are input in preparation for the solution runs. These steps complete preparation of the first model and it is ready to run.

Run the analysis program and at the run completion, assuming no errors have been reported, check the output listing file and scan the reflected listing of the input data parameters. Access the program's postprocessor and prepare a deformed shape displacement plot and study it for agreement with the applied boundary conditions and expected deformation. Examine the stress results through the graphic displays and compare these to boundary condition values and engineering calculations that were made with approximate equations. These steps provide a good check that the first model was done correctly and approximates the actual behavior. Thorough study of all the results provides further insight into the structural behavior.

Evaluation of the results from this first model will show where to refine the model to begin the convergence to an accurate solution. Regions within the model with high stress values and rapid variation as well as regions of low stress are selected for refinement. Reducing element size in regions of high stress values or rapid variation provides refinement. Increasing the element size in regions of low stress, if practical, provides refinement by keeping the number of system equations down. Convergence of results is very important to assure validity of the analysis.

For example, look at the results from two models of a quarter section of a tensile loaded bar with a center hole in Figure 1-6 and Figure 1-7. These figures are stress contour plots of the normal stress parallel to the applied load for two different models. The results in Figure 1-7 are from a refinement of the model that produced the results in Figure 1-6. The refinement raised the value of maximum stress predicted for the case. However, it will be shown in further study of this case in Chapter 5 that convergence was not yet reached.

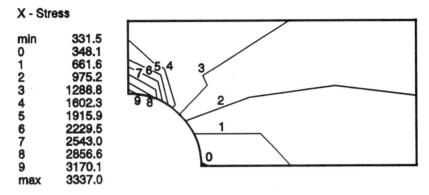

X - Stress

min	331.5
0	348.1
1	661.6
2	975.2
3	1288.8
4	1602.3
5	1915.9
6	2229.5
7	2543.0
8	2856.6
9	3170.1
max	3337.0

Figure 1-6. Stress in a Tensile Bar with a Hole Using Model A

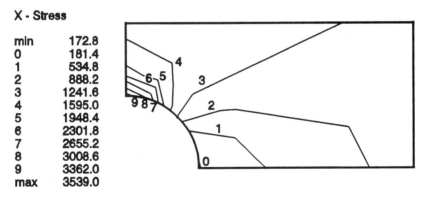

X - Stress

min	172.8
0	181.4
1	534.8
2	888.2
3	1241.6
4	1595.0
5	1948.4
6	2301.8
7	2655.2
8	3008.6
9	3362.0
max	3539.0

Figure 1-7. Stress in a Tensile Bar with a Hole Using Model B

A serious mistake would be made if only one model was analyzed with no further refinement. Both of the contour plots shown here look very reasonable and very believable. However, if the analyst quit after running the first model, the result would be in error by 23 percent! Even the second model produces an error of 19 percent. This illustrates that using a finite element program is no guarantee that the results will be accurate although the graphic display may be very convincing. Accurate analyses come about by applying good judgment and good technique to the practice of finite element analysis.

Running the second analysis with the refined mesh provides a second solution that you may compare with the first solution to check convergence. Examine the element-to-element variations for reasonable continuity in the second analysis. Compare the relative values between the two solutions, and then project or extrapolate to a better estimate of the actual solution. Judgment of these comparisons help decide on further refinements needed to reach the desired convergence.

Rate the final analysis by estimating the accuracy achieved and determining if the important criteria identified at the beginning of the analysis were satisfied. Repeat this cycle until you convince yourself of the solution validity.

The completion of the analysis leads to preparation of results for presentation and any required reporting. Keep in mind during this stage that the most significant results are of most interest and the presentation should primarily focus on those results. They will support or justify any decisions about the component design. It is also important to remember that the quality of the analysis is judged by others based upon the presentation of results. Therefore, good analysis work deserves a good presentation. After selecting the results for presentation, develop the report and make recommendations of actions to take.

This covers the significant stages in the process of an analysis. While it is not normally a formalized procedure, it shows the proper sequence in which to make the steps and identifies the importance of each stage. It is very important to conclude with a complete analysis that is accurate and provides the insight needed to decide on any design changes needed for the component or structure.

Problems

1.1-1.5 Using the one-dimensional spring element formulation develop the structure stiffness matrix and apply the indicated boundary conditions to get the resulting system equations for the assembly shown in the figures.

1.6 If the spring constants in Figure P1-2 are $k_1 = 100$, $k_2 = 200$, and $k_3 = 300$ lb/in with the load $F_2 = 150$ lb find the displacement of node 2.

1.7 If the spring constants in Figure P1-3 are $k_1 = k_2 = 1000$, $k_3 = 500$, and $k_4 = 200$ lb/in with loads $F_2 = 500$ and $F_3 = 300$ lb find the displacement of nodes 2 and 3 and the forces in each spring.

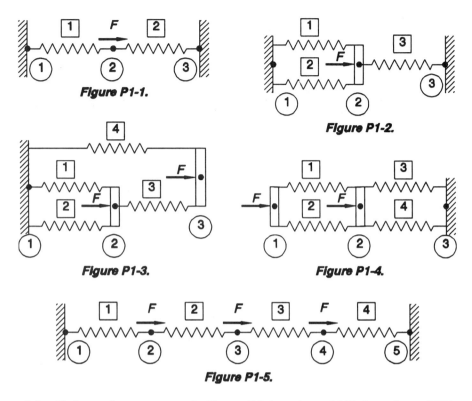

Figure P1-1.

Figure P1-2.

Figure P1-3.

Figure P1-4.

Figure P1-5.

1.8   If the spring constants in Figure P1-4 are k_1 = 2000, k_2 = k_3 = 4000, and k_4 = 10,000 N/m with loads F_1 = 500 and F_2 = 1000 N find the displacement of nodes 1 and 2 and the forces in each spring.

References

1.1   Zienkiewicz, O. C. and Cheung, Y. K., *The Finite Element Method in Structural and Continuum Mechanics*, McGraw-Hill, London, 1967.

1.2   Akin, J. E., *Finite Element Analysis for Undergraduates*, Academic Press, London, 1986.

1.3   Gallagher, R. H., *Finite Element Analysis - Fundamentals*, Prentice-Hall, Englewood Cliffs, New Jersey, 1975.

1.4   Huebner, K. H. and Thornton, E. A., *The Finite Element Method for Engineers*, John Wiley and Sons, New York, 1982.

1.5   Irons, B. and Ahmad, S., *Techniques of Finite Elements*, John Wiley and Sons, New York, 1980.

1.6   Irons, B. and Shrive, N., *Finite Element Primer*, John Wiley and Sons, New York, 1983.

1.7 Reddy, J. N., *An Introduction to the Finite Element Method*, McGraw-Hill, New York, 1984

1.8 Zienkiewicz, O. C. and Taylor, R. L., *The Finite Element Method, Volume 1 - Basic Formulation and Linear Problems*, Fourth Edition, McGraw-Hill, New York, 1989.

1.9 Cook, R. D., Malkus, D. S., and Plesha, M. E., *Concepts and Applications of Finite Element Analysis*, Third Edition, John Wiley and Sons, New York, 1989.

1.10 Fenner, D. N., *Engineering Stress Analysis: A Finite Element Approach with Fortran 77 Software*, John Wiley and Sons, New York, 1987.

1.11 Potts, J. F. and Oler, J. W., *Finite Element Applications with Microcomputers*, Prentice-Hall, Englewood Cliffs, New Jersey, 1989.

1.12 Ross, C. T. F., *Finite Element Methods in Structural Mechanics*, John Wiley and Sons, New York, 1985.

1.13 Stasa, F. L., *Applied Finite Element Analysis for Engineers*, Holt, Rinehart and Winston, New York, 1985.

1.14 Weaver, W. and Johnston, P. R., *Finite Elements for Structural Analysis*, Prentice-Hall, Englewood Cliffs, New Jersey, 1984.

1.15 Akin, J. E., *Application and Implementation of Finite Element Methods*, Academic Press, London, 1982.

1.16 Bathe, K. J. and Wilson, E. L., *Numerical Methods in Finite Element Analysis*, Prentice-Hall, Englewood Cliffs, New Jersey, 1976.

1.17 Bathe, K. J., *Finite Element Procedures in Engineering Analysis*, Prentice-Hall, Englewood Cliffs, New Jersey, 1981.

C H A P T E R 2
_____

STRUCTURES AND ELEMENTS

While all engineering structures are fully three-dimensional in volume, solution of many stress analysis problems are on simpler domains of lesser dimension. Often the simpler domains represent the three-dimensional member very well and there is no need to do a 3-D analysis. In other cases requiring a 3-D analysis, sometimes the engineer can approximate the behavior with a simpler domain and then use the insight gained from a simpler analysis to perform a good 3-D analysis. Classifications of engineering structural components include trusses, beams, plates, shells, plane solids, axisymmetric solids, and torsion bars. Each classification has its corresponding mechanics theory of behavior.

One of the primary steps in doing a finite element analysis is proper element selection to represent the structure behavior in the model. Element formulations follow the traditional structural classification and as such then provide many choices to model a particular structure. If the structure's geometry and loading fit a simpler class, then it becomes unnecessary to consider doing a fully three-dimensional analysis because we gain no higher solution accuracy.

This chapter describes the classes of engineering structures and introduces the typical corresponding finite element. It reviews the basic assumptions involved in these engineering structure classifications and relates to the methods for formulation of the elements. Later chapters will present details of the formulation of each element.

2.1 Trusses

The simplest type of engineering structure consists of members called *trusses*. The truss structure introduced in elementary mechanics interconnects truss members in such a manner that the structure can carry loads without kinematic collapse [2.1].   A truss member illustrated in Figure 2-1 is slender (its length is much larger than the cross section dimensions). The member is a two-force member and therefore only has capability to support tensile or compressive loads axially along the length. Members are joined by pins so no flexural loading transmits between members. The cross sectional dimensions and elastic properties of each member are constant along its length.

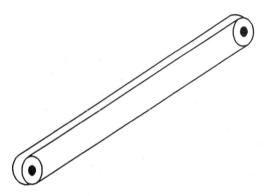

Figure 2-1. Truss Member

The members may interconnect in a 2-D or 3-D configuration in space depending on the support needed for the loading. The member is equivalent to a one-dimensional spring in that it has no stiffness against applied loads except those acting along the axis of the member. Therefore, the engineer or designer is responsible for setting up a kinematically stable configuration to support the applied loading.

The result of all the assumptions is that the displacement of material particles positioned along a member is a linear function of position. Define the strain in the member by

$$\epsilon = \frac{du}{dx} \tag{2.1}$$

where, ϵ is the axial strain, u is the axial displacement of any point along the length, and x is the coordinate along the length. If the displacement is a linear function of the length, then the strain is constant along the member. The stress relates to the strain through the one-dimensional Hooke's law

$$\sigma = E \epsilon \qquad (2.2)$$

where, σ is the axial stress, and E is the modulus of elasticity. Therefore, the stress along the member is also constant.

If the truss member were not uniform along its length, then it violates these assumptions. For example, if weight reducing holes drilled through the cross section at regular spacing existed along the length, then the stress distribution and strain distribution would be very nonuniform and at least 2-D in nature. However, to analyze the truss structure consisting of one or more of these members use an equivalent member. Do a two- or three-dimensional analysis on a repeating segment of the member subject to axial loads. Determine the stress and strain concentration factors and an equivalent modulus of elasticity to provide the correct stiffness in the truss structure model. The results from the truss structure model then provide the data to calculate the actual stress and strain in any member.

A two-dimensional truss structure appears in Figure 2-2. Each line in this figure represents a truss member. Theoretically the loads must apply to the connecting joints of members, while in practical applications distributed loading may occur along some members. In those cases a truss analysis may only be done by replacing the distributed loads with statically equivalent loads at the joints. The analysis of such a loaded member itself would have to be done using another class of structure, such as a beam or two-dimensional solid.

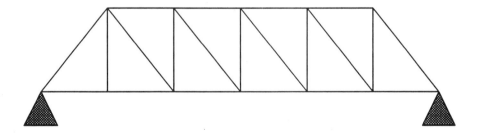

Figure 2-2. Truss Structure

A truss finite element is a line element connecting two points with stiffness equivalent to a one-dimensional spring. The finite element formulation uses the same structural assumptions used to classify a member as a truss. Therefore, the finite element formulation representing the truss is exact. In creating a finite element model of the structure in

Figure 2-2 each member becomes one element.   Since the element formulation is exact, it requires no member subdivision into more than one element.   In fact, subdividing the members into more than one element creates numerical problems.

2.2 Beams

Another structure class from strength of materials is a *beam* [2.2]. Beams are also members whose length is much greater than its transverse dimensions. However, they support loads applied lateral to the length. This causes bending or flexural response of the member. Elementary beam theory applies, if the member meets the conditions described by the following list of assumptions.

The major assumptions for the beam element are: (1) the member cross section is constant along the length of the member, (2) the cross section dimensions are small compared to the overall length, and (3) the stress and strain vary linearly across the section depth. The linear variation of stress and strain changes from tensile on one side of the member to compression on the other. This results in an internal moment to resist the external moment due to the applied external loading. A linear elastic analysis requires small displacements compared to the lateral physical dimensions and material properties that are linear elastic and constant along the length of the element.

Many simple beams are statically determinant allowing calculation of the normal stress directly from the applied loading without considering the deformation of the structure. However, in statically indeterminant cases, beam deflection influences the result. The assumption of small displacements and small curvatures of the beam under its applied loading allows us to describe the motion of all points by the lateral displacement and the slope of the displacement curve. The force and displacement components along a typical beam segment are shown in Figure 2-3. They are the transverse shear force, V, the moment, M, the lateral displacement, v, and the slope or rotation, ϕ, of the section at each end of the beam.

These components are all related through the differential equations of elementary beam theory. If the beam segment has no distributed loading, the equation for the displacement shape is a cubic polynomial in x, where x is the coordinate along the length of the beam. This satisfies the beam differential equation and forms the basis for the finite element formulation based on displacements.

Loading of a straight beam causes the longitudinal neutral axis of the beam to bend into the elastic curve. The elastic curve must satisfy the differential equation (2.3) where q is the distributed load intensity.

On all sections of a beam without any distributed load, q is zero and the

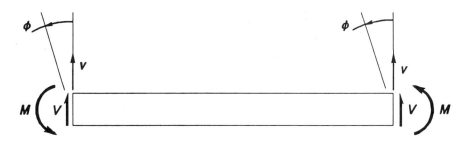

Figure 2-3. Beam Section

$$EI\frac{d^4v}{dx^4} = q \qquad (2.3)$$

solution to the homogeneous equation is a cubic polynomial in x. The constant coefficients of its terms are calculated to match the boundary conditions at each end of the beam section.

The longitudinal strain is proportional to the distance from the neutral axis and the second derivative of the elastic curve. Expressed in equation form it is

$$\epsilon = y\frac{d^2v}{dx^2} \qquad (2.4)$$

where y is the coordinate of any point on the beam cross section from the neutral axis. There is also a shear strain component that exists due to the transverse shear stress created by the transverse shear force, V, acting on the beam cross section. However, for beams that satisfy the assumptions of elementary beam theory its size is negligible.

There is little advantage to a finite element approach for simple beams. However, for complex three-dimensional interconnected beam structures the finite element method makes an intractable problem much simpler. The basis of finite element formulation is the cubic displacement form and therefore provides no greater accuracy than conventional beam theory. It is an exact formulation for those beam spans that have no distributed load, but it is only approximate for segments with a distributed load. Therefore, in a typical model, representation of segments without distributed load is exact and there is no need for subdivision within that segment. All beam segments with a distributed load require element subdivision and the application of an equivalent work load set at the nodes to converge to an accurate solution.

Some further features of the finite element formulation add to its ease of use in solving complex beam problems. They include the ability to handle arbitrary orientations in three-dimensional space, to account for cross section changes from one element to the next, and to add supporting structure stiffness into the model.

2.3 Two-Dimensional Solids

Classification as a *two-dimensional solid* occurs when the geometry definition lies in a plane and the applied loads lie in the same plane. There are two general cases of two-dimensional problems illustrated in Figure 2-4 called plane stress and plane strain [2.3]. The plane stress condition occurs for structures with a small thickness compared with its in-plane dimensions. In this case the stress components associated with the out-of-plane coordinate are all zero. The plane strain condition occurs whenever the thickness becomes very large compared with its in-plane dimensions. In this case, the strains in the out-of-plane direction are all zero.

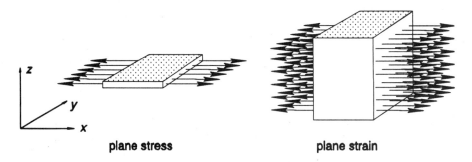

plane stress   plane strain

Figure 2-4. Two-Dimensional Solid

The theoretical definition calls for the thickness in plane stress to approach zero, while in plane strain the thickness approaches infinity. If the thickness is finite and about the same size as the in-plane dimensions, then it is a three-dimensional problem. However, in practice if only in-plane loads exist, then the plane stress and plane strain solutions usually bound the true three-dimensional solution. The plane stress and plane strain solutions are usually similar and in some cases are the same.

Equations derived in theory of elasticity govern the solution to problems in two-dimensional solids. These equations relate displacement, strain, and stress components.

There are two displacement components describing the movement of any point in the plane of the two-dimensional structure. Locating the object in a x,y coordinate system produces the components u in the x direction and v in the y direction. The strain components that exist in the plane of solution are ϵ_x, ϵ_y, and γ_{xy}.

The strain-displacement relationships are given in equation (2.5).

$$\epsilon_x = \frac{\partial u}{\partial x}$$

$$\epsilon_y = \frac{\partial v}{\partial y} \qquad (2.5)$$

$$\gamma_{xy} = \frac{\partial u}{\partial y} + \frac{\partial v}{\partial x}$$

The stress components corresponding to these strains are σ_x, σ_y, and τ_{xy}. In plane stress the components σ_z, τ_{xz}, and τ_{yz} are all zero. The stress-strain relations for plane stress are given in equation (2.6)

$$\begin{Bmatrix} \sigma_x \\ \sigma_y \\ \tau_{xy} \end{Bmatrix} = \frac{E}{(1-\nu^2)} \begin{bmatrix} 1 & \nu & 0 \\ \nu & 1 & 0 \\ 0 & 0 & \frac{1-\nu}{2} \end{bmatrix} \begin{Bmatrix} \epsilon_x \\ \epsilon_y \\ \gamma_{xy} \end{Bmatrix} \qquad (2.6)$$

where σ_x and σ_y are normal stress components, τ_{xy} is the shear stress, E is the modulus of elasticity, and v is the poisson's ratio.

In plane strain the out-of-plane strain components ϵ_z, γ_{xz}, and γ_{yz} are zero, the stress components σ_z, τ_{xz}, and τ_{yz} are nonzero, and σ_z may become a significant value. We may compute the value of σ_z in the plane strain case directly from the in-plane components of the two-dimensional solution provided the plane strain condition of $\epsilon_z = 0$ is met.

The stress-strain relations for plane strain are given in equations (2.7).

$$\begin{Bmatrix} \sigma_x \\ \sigma_y \\ \tau_{xy} \end{Bmatrix} = \frac{E}{(1+\nu)(1-2\nu)} \begin{bmatrix} 1-\nu & \nu & 0 \\ \nu & 1-\nu & 0 \\ 0 & 0 & \frac{1-2\nu}{2} \end{bmatrix} \begin{Bmatrix} \epsilon_x \\ \epsilon_y \\ \gamma_{xy} \end{Bmatrix} \qquad (2.7)$$

More common in plane strain is a condition called *generalized plane strain* that requires the strain component ϵ_z to be uniform but not zero. In this case the value of σ_z depends on the value of applied load in the z direction to calculate the uniform ϵ_z. For example, if there is no z load applied to the structure, the average value of σ_z over the cross section must be zero.

The finite element formulation approximates the displacement solution within an element by a simple functional form relating to values at the node points. By assuming this function for the displacement, we derive the element stiffness matrix relating nodal displacements to nodal forces. The simplest displacement approximation leads to an element formulation that has constant stress and strain values within the element. The values may differ from one element to the next to solve problems where the stress and strain vary across the region. However, the resulting approximation requires subdivision of the region into several elements to make a reasonable approximate solution.

For example, the simple cantilever beam illustrated in Figure 2-5, if modeled by 2-D elements would require element subdivision along its length and its height to approximate the known variations in stress and strain that occur. The earliest two-dimensional elements were triangular shaped. However, most analyses are now done with quadrilateral elements because linear triangle elements do not work well in most cases.

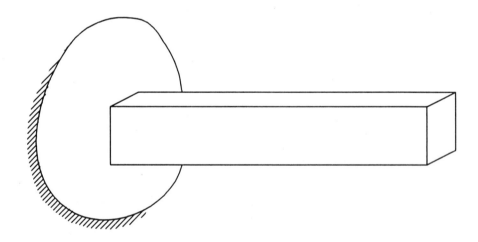

Figure 2-5. Cantilevered Beam

A finite element mesh of the cantilever beam is shown in Figure 2-6. This illustrates the type of subdivision required to develop reasonable estimates of stresses throughout the beam.

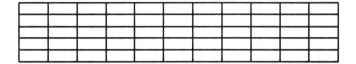

Figure 2-6. Two-Dimensional Finite Element Model of Cantilever Beam

2.4 Three-Dimensional Solids

A *three-dimensional solid* classification covers all structures, and thus may have any geometry describable in 3-D space and any arbitrary loading. However, this classification is reserved for use only if no simpler classification exists. The theory of elasticity provides the governing equations [2.3] that consist of 12 independent partial differential equations, 6 stress-strain relations with any prescribed force and displacement boundary conditions. Obviously, very few closed form solutions exist.

The strain-displacement and stress-strain relations presented here will help familiarize the reader with some of the mechanics that go into a displacement based finite element formulation.

The strain-displacement relations are shown in equations (2.8).

$$\epsilon_x = \frac{\partial u}{\partial x}$$

$$\epsilon_y = \frac{\partial v}{\partial y}$$

$$\epsilon_z = \frac{\partial w}{\partial z}$$

$$\gamma_{xy} = \frac{\partial u}{\partial y} + \frac{\partial v}{\partial x} \tag{2.8}$$

$$\gamma_{xz} = \frac{\partial u}{\partial z} + \frac{\partial w}{\partial x}$$

$$\gamma_{yz} = \frac{\partial v}{\partial z} + \frac{\partial w}{\partial y}$$

The stress-strain relations for an isotropic homogeneous material are given in equations (2.9).

The other governing equations are the force equilibrium equations and compatibility equations. These equations with appropriate boundary conditions are all satisfied in an elasticity solution.

$$\begin{Bmatrix} \sigma_x \\ \sigma_y \\ \sigma_z \\ \tau_{xy} \\ \tau_{xz} \\ \tau_{yz} \end{Bmatrix} = \frac{E}{(1+\nu)(1-2\nu)} \begin{bmatrix} 1-\nu & \nu & \nu & 0 & 0 & 0 \\ \nu & 1-\nu & \nu & 0 & 0 & 0 \\ \nu & \nu & 1-\nu & 0 & 0 & 0 \\ 0 & 0 & 0 & \frac{1-2\nu}{2} & 0 & 0 \\ 0 & 0 & 0 & 0 & \frac{1-2\nu}{2} & 0 \\ 0 & 0 & 0 & 0 & 0 & \frac{1-2\nu}{2} \end{bmatrix} \begin{Bmatrix} \epsilon_x \\ \epsilon_y \\ \epsilon_z \\ \gamma_{xy} \\ \gamma_{xz} \\ \gamma_{yz} \end{Bmatrix} \quad (2.9)$$

The simplest finite element for 3-D analysis is a four-node tetrahedron, but they perform very poorly and require many elements to get adequate solutions. The most commonly used element is a hexahedron with eight corner nodes. These are sketched in Figure 2-7. Again within the element a simple function for the material displacement is chosen. Representing allowable variations in results throughout the volume requires structure subdivision into elements in all directions.

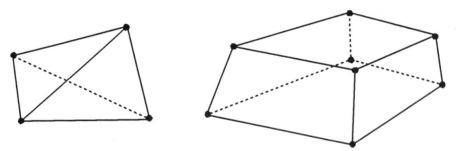

Figure 2-7. Three-Dimensional Solid Finite Elements

This becomes a very large task with large associated costs of computing and human resources. Although these costs are dropping with improved software and hardware, it is better to develop some 2-D approximation to the actual problem and solve it first. Then we may more efficiently model the 3-D problem and thus reduce the effort.

2.5 Axisymmetric Solids

If a three-dimensional body of revolution, an axisymmetric geometry, has applied loads that are also axisymmetric then the classification is an *axisymmetric solid structure* [2.3]. These conditions allow reduction of the three-dimensional elasticity theory. There are many practical applications of axisymmetric structures including thick-wall pressure vessels both cylindrical and spherical as well as rotating objects such as flywheels and turbine disks. An axisymmetric body with a finite element mesh of the cross section is given in Figure 2-8.

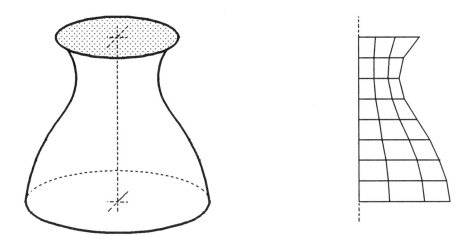

Figure 2-8. Axisymmetric Solid

A cylindrical, r-θ-z, coordinate system is usually chosen for axisymmetric problems with r pointing along the radius from the axis, θ the circumferential direction, and z the axial coordinate. The axisymmetric assumption requires that the solution be independent of the θ position. The displacement of any material particle then lies in the r-z plane. If $u, v,$ and w are the three respective displacement components for this coordinate system, then $v = 0$. This allows much simplification in the solution and it becomes very close to the two-dimensional problem.

There are four strain components defined by the two displacement components. The two shear strain components acting in the circumferential direction must equal zero. The strain-displacement relations are shown in equations (2.10).

$$\epsilon_r = \frac{\partial u}{\partial r}$$

$$\epsilon_\theta = \frac{u}{r}$$

$$\epsilon_z = \frac{\partial w}{\partial z}$$ (2.10)

$$\gamma_{rz} = \frac{\partial u}{\partial z} + \frac{\partial w}{\partial r}$$

Since there are only four strain components the stress-strain relations reduce to

$$
\begin{Bmatrix} \sigma_r \\ \sigma_\theta \\ \sigma_z \\ \tau_{rz} \end{Bmatrix} = \frac{E}{(1+\nu)(1-2\nu)} \begin{bmatrix} 1-\nu & \nu & \nu & 0 \\ \nu & 1-\nu & \nu & 0 \\ \nu & \nu & 1-\nu & 0 \\ 0 & 0 & 0 & \frac{1-2\nu}{2} \end{bmatrix} \begin{Bmatrix} \epsilon_r \\ \epsilon_\theta \\ \epsilon_z \\ \gamma_{rz} \end{Bmatrix}
$$ (2.11)

The finite element formulation approximates the displacement solution for axisymmetric solids similar to the 2-D plane case. The degrees-of-freedom involved in the element formulation become exactly the same and the element stiffness matrix is then the same size as the 2-D problem. The differences come about by the addition of terms to account for the hoop direction components of stress and strain although there are no hoop displacement or force components. In practice, inclusion of these terms in a finite element program formulation is very simple. Usually, the axisymmetric element is an option available within the general class of 2-D elements.

2.6 Torsion Bars and Shafts

Sometimes a twisting or torsional load acts on a machine component or structure. If the geometry of the structure is a simple circular cylinder with the torsion loads applied at each end, the solution comes from elementary torsion formulas [2.2]. If the structure is a noncircular cylinder, then the solution becomes more involved, but it is still simpler than a full

3-D problem.   If the structure also has geometry variations along the length, then the problem may become 3-D in nature.

Two cases require less than a full 3-D approach.   One of these, sketched in Figure 2-9, has a noncircular cross section, but the body is still cylindrical with no variation of the section along the length.   The other case, sketched in Figure 2-10, has a circular cross section with a varying diameter along the length.   In the first case, the cross section and length define the geometry.   In the second case, the geometry is an axisymmetric body of revolution.   Machinery parts of these types abound in mechanical engineering applications for torsion bars and shafts.   Some few closed form solutions exist for the simpler shapes, but in most cases they do not exist.

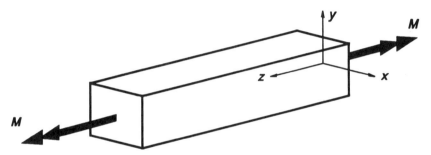

Figure 2-9. Torsion of Non-Circular Cross Sections

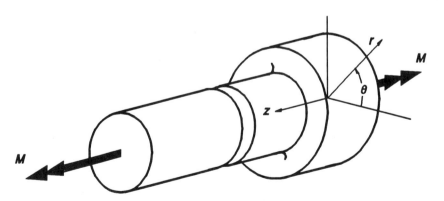

Figure 2-10. Torsion of Axisymmetric Bodies

Noncircular prismatic bars with application of a torsional load experience an out-of-plane displacement or warping of the cross section.   Saint-

Venant, using a technique called the semi-inverse method, formulated the general solution to this elasticity problem in terms of a warping function. The solution is in many texts on theory of elasticity [2.3].

A brief description of this solution follows. Set up the coordinate system for the prismatic bar so the cross section lies in the x-y plane and z runs along the length of the bar. As with circular cross sections, Saint-Venant assumed that a straight torsion member has an axis of twist. Each cross section then rotates about this axis in response to torque loading. The amount or angle of relative rotation is linearly proportional to the distance, z, between the applied and reaction torque. Define θ as the angle of twist per unit length, and then the rotation angle of any cross section β is given by

$$\beta = \theta z. \tag{2.12}$$

The displacement components in the x-y cross section plane are then given by

$$\begin{aligned} u &= -\theta zy \\ v &= \theta zx. \end{aligned} \tag{2.13}$$

The out-of-plane displacement is a function of x and y over the cross section but constant for all cross sections. Its magnitude is

$$w = \theta \psi \tag{2.14}$$

where, ψ is the warping function. Application of the theory of elasticity will show that the warping function satisfies the Laplace equation

$$\nabla^2 \psi = 0 \tag{2.15}$$

over the cross section area. Use of these displacement components in the three-dimensional strain-displacement relations show that the strain components are given by equations (2.16).

$$\epsilon_x = \epsilon_y = \epsilon_z = \gamma_{xy} = 0$$

$$\gamma_{zx} = \theta \left(\frac{\partial \psi}{\partial x} - y \right) \tag{2.16}$$

$$\gamma_{zy} = \theta \left(\frac{\partial \psi}{\partial y} + x \right)$$

With only two strain components, the stress-strain relations reduce to equations (2.17). Here, G is the shear modulus of elasticity and is related to Young's modulus and Poisson's ratio by equation (2.18).

$$
\begin{Bmatrix} \tau_{xz} \\ \tau_{yz} \end{Bmatrix} = \begin{bmatrix} G & 0 \\ 0 & G \end{bmatrix} \begin{Bmatrix} \gamma_{xz} \\ \gamma_{yz} \end{Bmatrix}
\tag{2.17}
$$

$$
G = \frac{E}{2(1+\nu)}
\tag{2.18}
$$

The finite element method approximates the warping function over the region of an element by a simple function with coefficients related to node point values of the element. So in the finite element solution, the unknowns are the values of the warping function at all the nodes. Unfortunately, most finite element codes do not provide a specific element in their element library to model this torsion problem. However, if a conduction heat transfer element is available, we can solve this torsion problem by proper interpretation of variables.

Use of the semi-inverse method applied to the equations of elasticity [2.3] also can solve for the torsion of axisymmetric members with a varying diameter. In this case, circumferential displacement of hoop fibers relates to a twisting function acting over the section of revolution of the axisymmetric body.

The cylindrical coordinate system is more convenient for this axisymmetric problem with r-θ-z representing the radial, circumferential, and axial directions respectively. The elasticity solution for this problem assumes that during torsion only the circumferential, v, displacement component is nonzero. Symmetry requires that its value is independent of θ, but it varies over the r-z plane. A twisting function, ϕ, is defined by equation (2.19).

$$
\phi = \frac{v}{r}
\tag{2.19}
$$

Use of the three-dimensional strain-displacement relations expressed in cylindrical coordinates shows there are only two nonzero strain components. They are given in equations (2.20). They have corresponding shear stress components.

Developing a finite element then involves approximating the twisting function within an element region by a simple function in terms of node

$$\gamma_{r\theta} = \frac{\partial v}{\partial r} - \frac{v}{r}$$

$$\gamma_{z\theta} = \frac{\partial v}{\partial z}$$

(2.20)

values. So in the finite element solution the unknowns are the values of the twisting function at all the nodes. Again most finite element programs exclude a specific element for this case unless the program can apply torsion loading to its axisymmetric solid element. It also must have corresponding output of the out-of-plane shear stress components.

2.7 Plates

Plates are another class of engineering structure [2.4]. The plate geometry lies in a plane, but the loads act normal to the plane as illustrated in Figure 2-11. The plate behavior is similar to a beam in that loads are supported by flexure. In a plate, the lateral depth dimension is about the same size as its length which are both very large compared to its thickness. Since the thickness is much smaller than the other dimensions, then assumptions that are similar to beam theory apply to the strain and stress distribution through the thickness.

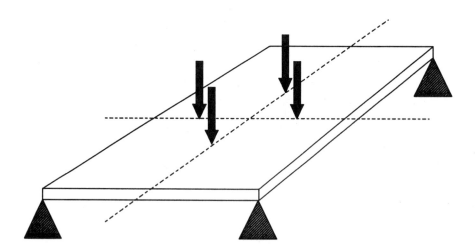

Figure 2-11. Plate Structure

In plates, flexural moments will exist in the material in both in-plane directions at any point. This generates a two-dimensional state of stress very similar to a plane stress condition. However, the components vary linearly through the thickness from tension on one face to compression on the opposite face. The requirements for a linear elastic analysis to apply and the assumptions associated with plate behavior include the following:

1. The deflection of the midsurface of the plate lateral to the plane is less than the thickness of the plate. This makes the slope of the deflected surface much less than unity.

2. Straight lines that are normal to the midsurface before loading remain straight and normal to the midsurface after loading. This makes the shear strain components γ_{xz} and γ_{yz} negligible.

3. There is no in-plane normal strain component at the midsurface of the plate implying that no stretching or contracting of the midplane occurs during loading of the plate.

4. The stress component σ_z normal to the plate surface and the shear components τ_{xz} and τ_{yz} are negligible.

If the plate lies in the x-y plane, then the displacement component w defines the displacement of the midsurface. The midsurface has no u or v displacement component. However, the material points above and below the midsurface do have u and v displacements. This is explained using Figure 2-12.

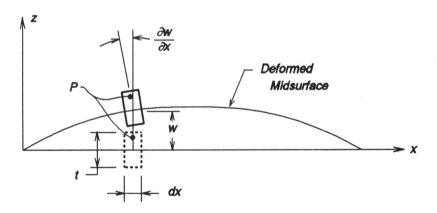

Figure 2-12. Plate Deformations

From an observation point along the y axis, when a differential slice of the plate takes the deformed position w it has a slope of $\partial w/\partial x$. Any material point, P, off the midsurface then has a u displacement given by equation (2.21). Similarly, looking at the y-z plane yields equation (2.22). The strain-displacement relations are formed in equation (2.23).

$$u = -z\frac{\partial w}{\partial x} \tag{2.21}$$

$$v = -z\frac{\partial w}{\partial y} \tag{2.22}$$

$$
\begin{aligned}
\epsilon_x &= \frac{\partial u}{\partial x} = -z\frac{\partial^2 w}{\partial x^2} \\
\epsilon_y &= \frac{\partial v}{\partial y} = -z\frac{\partial^2 w}{\partial y^2} \\
\gamma_{xy} &= \frac{\partial u}{\partial y} + \frac{\partial v}{\partial x} = -2z\frac{\partial^2 w}{\partial x \partial y}
\end{aligned}
\tag{2.23}
$$

The stress-strain relations are the same as those for plane stress presented earlier. In thin plate theory, these equations combine with the equilibrium equations to produce the governing partial differential equation

$$\nabla^4 w = \frac{q}{D} \tag{2.24}$$

where, q is the surface pressure load and D, the flexural rigidity, is given by

$$D = \frac{Et^3}{12(1-\nu^2)} \, . \tag{2.25}$$

The formulation of the finite element approach to plates follows the approach to beams in describing the complete geometry of the structure by its in-plane geometry and thickness. The motion of the material above and below the midsurface relates to the midsurface motion. The displacement formulation describes the displacement of any material point by its u, v, w displacement components. The relationships and assumptions above show that the lateral displacement w is independent while u and v are determined by the derivative of the lateral displacement taken with respect to x and y. Therefore, the assumed displacement function within an element must approximate the lateral displacement w and the two partial derivatives. This requires the minimum degrees-of-freedom at the node points to be w, $\partial w/\partial x$, and $\partial w/\partial y$. Using a well formulated element, we subdivide the in-plane geometry into elements to find the solution over the region of the plane.

2.8 Shells

Structural members that are similar to plates but defined on a curved surface rather than a plane are *shells* [2.5]. The surface geometry and thickness describe the member as in Figure 2-13. The thickness is very small relative to its remaining dimensions. There are many practical examples of shell structures including light bulbs, airplanes, low pressure tanks and many other types of containers.

Figure 2-13. Shell Structure

Shells will support loads in much the same way that plates do with the addition of a sizable in-plane support. Therefore, in this case, there is flexure of the shell thickness coupled with membrane strain in the tangent plane to the shell. There are several theories describing the mechanical behavior of shells. Many of these produce governing equations that are just as difficult or more difficult to solve than three-dimensional elasticity.

There are many shell theories and therefore more associated strain-displacement relations than can be effectively listed in this short summary of shell behavior. The membrane shell theory assumes the shell thickness is so small there is negligible load carrying capacity due to shell bending. In this case, only the displacements of the midsurface matter. However, most shell theories include both membrane and bending contributions so, unlike plate theory, the midsurface displacements tangent to the surface also contribute to the deformation.

There are many formulations for special types of shells like cylinders or shallow segments because the general shell theory is so complex. In all of them though, the displacement of any material point is determined by combining the midsurface values and the local flexure rotations. Because the shell is thin, the stress-strain relations for plane stress also apply to shells.

The finite element formulation may be done several different ways depending on which theory the formulator uses to approximate the shell behavior. In all cases, the nodes have a minimum of three displacement components and two rotation components in the surface tangent plane as degrees-of-freedom in the model. The analyst should realize that the finite element solution for general shells is difficult to interpret and prone to large error if the formulation or shell theory used does not match the problem.

2.9 Closure

This section concludes an overview of the traditional classes of engineering structures and components. The approach normally used to formulate a finite element that fits the structural classification was indicated. The following sections describe in more detail some aspects of the element formulation and the application of that formulation to model properly the engineering structure.

Problems

2.1   Select at least four parts of a bicycle, and classify each into the mechanical structure (truss, beam, plate, etc.) that will adequately characterize its behavior. Describe any approximations you made in the classification and what limitations may exist in any solution using the chosen classification.

2.2   Select at least four parts of an automobile, and classify each into the mechanical structure (truss, beam, plate, etc.) that will adequately characterize its behavior. Describe any approximations you made in the classification and what limitations may exist in any solution using the chosen classification.

2.3   Classify the structural members shown in Figure P2-3 below and
state what type of finite element to select from the element library
of a finite element program to model the member.

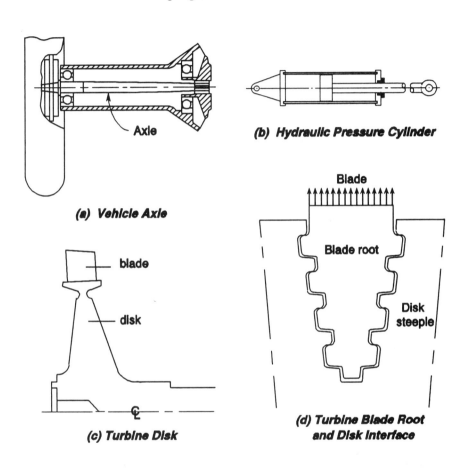

(a) *Vehicle Axle*

(b) *Hydraulic Pressure Cylinder*

(c) *Turbine Disk*

(d) *Turbine Blade Root and Disk Interface*

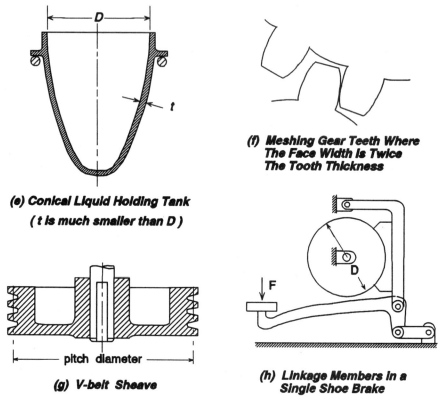

(e) Conical Liquid Holding Tank
(t is much smaller than D)

(f) Meshing Gear Teeth Where
The Face Width Is Twice
The Tooth Thickness

(g) V-belt Sheave

(h) Linkage Members In a
Single Shoe Brake

Figure P2 - 3.

References

2.1   Shames, I. H., *Engineering Mechanics, Volume 1, Statics*, Prentice-Hall, Inc., Englewood Cliffs, New Jersey, 1980.

2.2   Popov, E. P., *Introduction to Mechanics of Solids*, Prentice-Hall, Inc., Englewood Cliffs, New Jersey, 1968.

2.3   Timoshenko, S. and Goodier, J. N., *Theory of Elasticity*, McGraw-Hill, New York, 1951.

2.4   Boresi, A. P. and Sidebottom, O. M., *Advanced Mechanics of Materials*, John Wiley and Sons, New York, 1985.

2.5   Flugge, W., *Stresses in Shells*, Springer-Verlag, Berlin, 1962.

C H A P T E R 3

TRUSSES

The primary focus of this text is on the aspects of finite element analysis that are more important to the user than the formulator or programmer. However, for the user to employ the method effectively he or she must have some understanding of the element formulations as well as some of the computational aspects of the programming. Therefore, the next several chapters begin by looking at the element formulation for a given structural behavior class before proceeding to model development and the proper modeling approach. Most finite elements develop from use of an assumed displacement approximation, therefore the elements presented will all deal with assumed displacement formulations.

3.1 Direct Element Formulation

This section presents the direct physical formulation of a truss element and its spatial orientation to solve two-dimensional frameworks. A member of a truss structure is like a one-dimensional spring. The member has a length substantially larger than its transverse dimensions, and it has a pinned connection to other members which eliminates all loads other than axial load along the member length. It usually has a constant cross section area and modulus of elasticity along its length. The stiffness is then

$$k = \frac{AE}{L} \tag{3.1}$$

where, A is the cross section area, E is the modulus of elasticity, and L is the member length.

A one-dimensional truss element then has an element formulation identical to the one-dimensional spring given in equation (1.2). This is its element stiffness matrix for one-dimensional displacement and loading along the axis of the member. For member positioning in a two-dimensional space as illustrated in Figure 3-1, each node has two components of displacement u and v and two components of force p and q. This leads to a set of element equations with an element stiffness matrix of size 4 by 4.

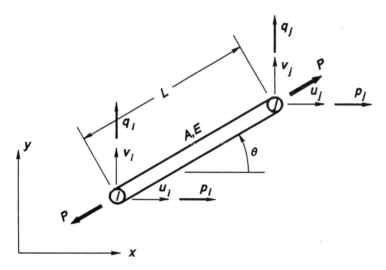

Figure 3-1. Two-Dimensional Truss Element

Derivation of the two-dimensional element stiffness matrix comes through coordinate transformations [3.1], but first we expand the one-dimensional stiffness matrix to two-dimensions with the member lying along the x axis. Assuming an order of components and equations of u and v at node i followed by u and v at node j the element equations are written in equation (3.2).

$$\begin{bmatrix} k & 0 & -k & 0 \\ 0 & 0 & 0 & 0 \\ -k & 0 & k & 0 \\ 0 & 0 & 0 & 0 \end{bmatrix} \begin{Bmatrix} u_i \\ v_i \\ u_j \\ v_j \end{Bmatrix} = \begin{Bmatrix} -p_i \\ -q_i \\ -p_j \\ -q_j \end{Bmatrix} \qquad (3.2)$$

Notice that the terms relating displacement and force in the x direction

are the spring constant of the member and the terms relating displacement
and force in the y direction are zero. A linear analysis always assumes
that the displacements are much smaller than the overall geometry of the
structure, therefore, the stiffness is based upon the undeformed configuration. In this case, if we consider a vertical displacement component at one
of the nodes, since it is a motion perpendicular to the line of the member,
no vertical force results because there is no axial stretch relative to the
undeformed configuration.

This formulation represents the element stiffness matrix in a local
element coordinate system that is aligned with the element axis.   To
position the element at an arbitrary angle, θ, from the x coordinate axis,
we perform a transformation of coordinate systems to derive the element
stiffness matrix in the x,y global coordinate system.   In the system of
equations, the displacements and forces are both vectors so they transform
through standard vector transformations. The displacement components
in global coordinates relate to local components through equation (3.3).

$$\{d\} = [T]\{d'\} \tag{3.3}$$

Here, $\{d'\}$ are the global displacement components, $[T]$ is the transformation matrix, and $\{d\}$ are the local element coordinate displacement
components.

The transformation matrix is given by equation (3.4).

$$[T] = \begin{bmatrix} c & s & 0 & 0 \\ -s & c & 0 & 0 \\ 0 & 0 & c & s \\ 0 & 0 & -s & c \end{bmatrix} \tag{3.4}$$

Here, s is the $\sin \theta$, and c is the $\cos \theta$. Similarly, the force components in
the global coordinate system are given by

$$\{f\} = [T]\{f'\} . \tag{3.5}$$

The element stiffness matrix in the local coordinate system is defined in
matrix notation from equation (3.2) by

$$[k]\{d\} = \{f\} . \tag{3.6}$$

Making the substitutions for $\{d\}$ and $\{f\}$ given above yields

$$[k][T]\{d'\} = [T]\{f'\} . \tag{3.7}$$

The transformation matrix is an orthogonal matrix meaning that

$$[T]^{-1} = [T]^T . \tag{3.8}$$

Therefore multiplying equation (3.7) by $[T]^T$ produces

$$[T]^T [k][T]\{d'\} = \{f'\} \tag{3.9}$$

which makes

$$[k'] = [T]^T [k][T] = k \begin{bmatrix} c^2 & cs & -c^2 & -cs \\ cs & s^2 & -cs & -s^2 \\ -c^2 & -cs & c^2 & cs \\ -cs & -s^2 & cs & s^2 \end{bmatrix} \tag{3.10}$$

The use of this element formulation and equation assembly is shown through the example truss structure pictured in Figure 3-2. The elements and nodes are numbered, and load and boundary conditions are shown. The structure equations are

$$[K]\{D\} = \{F\} \tag{3.11}$$

where, $[K]$ is the structure stiffness matrix, $\{D\}$ is the node displacement vector, and $\{F\}$ is the applied load vector.

These equations come from applying the conditions of equilibrium to all the nodes by setting the summation of internal forces equal to the applied forces. The internal forces are given by the product of each element stiffness matrix with its node displacements. This yields equation (3.12),

$$[k]\{d\}|_1 + [k]\{d\}|_2 + [k]\{d\}|_3 = \{F\} \tag{3.12}$$

where the subscripts refer to the numbered elements. If the displacement vector in each term of the equation above was identical, then we could

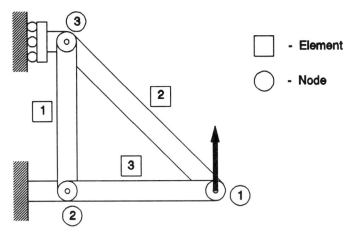

Figure 3-2. Example Truss Structure

factor it out and add the stiffness matrices term-by-term to produce the structure stiffness matrix.

The displacement vector for each element must then expand to include all the structure degrees-of-freedom not just the ones associated with a given element. In order for the matrix equation to be correct, a corresponding expansion of the element stiffness matrix must accompany the expansion of the displacement vector. It expands to the size of the structure stiffness matrix which in this example becomes a 6 by 6 matrix. The expansion simply adds rows and columns of zeroes to each element stiffness matrix corresponding to the additional structure degrees-of-freedom unused in the given element [3.1].

Applying this approach, the stiffness matrix for element 1 in the example results from using equation (3.10) with a θ value of 90 degrees. Rows and columns of zeroes fill in equations and positions involving u_1 and v_1 as shown in equation (3.13).

$$[k]_1 = \begin{bmatrix} 0 & 0 & 0 & 0 \\ 0 & k_1 & 0 & -k_1 \\ 0 & 0 & 0 & 0 \\ 0 & -k_1 & 0 & k_1 \end{bmatrix} = \begin{bmatrix} 0 & 0 & 0 & 0 & 0 & 0 \\ 0 & 0 & 0 & 0 & 0 & 0 \\ 0 & 0 & 0 & 0 & 0 & 0 \\ 0 & 0 & 0 & k_1 & 0 & -k_1 \\ 0 & 0 & 0 & 0 & 0 & 0 \\ 0 & 0 & 0 & -k_1 & 0 & k_1 \end{bmatrix} \quad (3.13)$$

Similarly, the matrix for element 2 with θ equal to 135 degrees and rows and columns filled in equations and positions involving u_2 and v_2 results in equation (3.14).

$$[k]_2 = \begin{bmatrix} .5k_2 & -.5k_2 & 0 & 0 & -.5k_2 & .5k_2 \\ -.5k_2 & .5k_2 & 0 & 0 & .5k_2 & -.5k_2 \\ 0 & 0 & 0 & 0 & 0 & 0 \\ 0 & 0 & 0 & 0 & 0 & 0 \\ -.5k_2 & .5k_2 & 0 & 0 & .5k_2 & -.5k_2 \\ .5k_2 & -.5k_2 & 0 & 0 & -.5k_2 & .5k_2 \end{bmatrix} \quad (3.14)$$

Finally for element 3, θ is 0 degrees and u_3 and v_3 are the additional degrees-of-freedom in equation (3.15).

$$[k]_3 = \begin{bmatrix} k_3 & 0 & -k_3 & 0 & 0 & 0 \\ 0 & 0 & 0 & 0 & 0 & 0 \\ -k_3 & 0 & k_3 & 0 & 0 & 0 \\ 0 & 0 & 0 & 0 & 0 & 0 \\ 0 & 0 & 0 & 0 & 0 & 0 \\ 0 & 0 & 0 & 0 & 0 & 0 \end{bmatrix} \quad (3.15)$$

The summations of equation (3.12) are now carried out by adding the expanded element stiffness matrices term by term. The resulting structure stiffness matrix is in equation (3.16).

$$[K] = \begin{bmatrix} .5k_2+k_3 & -.5k_2 & -k_3 & 0 & -.5k_2 & .5k_2 \\ -.5k_2 & .5k_2 & 0 & 0 & .5k_2 & -.5k_2 \\ -k_3 & 0 & k_3 & 0 & 0 & 0 \\ 0 & 0 & 0 & k_1 & 0 & -k_1 \\ -.5k_2 & .5k_2 & 0 & 0 & .5k_2 & -.5k_2 \\ .5k_2 & -.5k_2 & 0 & -k_1 & -.5k_2 & .5k_2+k_1 \end{bmatrix} \quad (3.16)$$

Before solving the equations, apply the displacement boundary conditions. In this example the boundary conditions have components u_2, v_2, and u_3 equal to zero. Applying these to the system equations zeroes the third, fourth, and fifth columns of the structure stiffness matrix. This leaves six equations with three unknown displacements. In most problems, the reaction forces in equations 3, 4, and 5 are also unknown so choose equations 1, 2, and 6 to solve for the displacement components, u_1, v_1, and v_3.

We get the solution by finding the inverse of the remaining 3 by 3 stiffness matrix. Multiplying the inverse with the load vector yields the displacements. After finding the displacements, calculate the element forces by use of the element equations.

This concludes the example and demonstration of two-dimensional truss formulation. Of course, it extends easily to three-dimensions. The assembly of equations by expanding the element stiffness matrix to structure size is useful for explanation of the process but impractical for a large number of system equations. In computer programs the algorithm only needs to place the terms of the element stiffness matrix in the correct position in the structure stiffness matrix. This is easy to accomplish since the structure equation sequence correlates to the node number. Elements are defined by node numbers, thus providing the direct correlation for positioning the terms.

3.2 Element Formulation by Virtual Work

In this section the virtual work principle is used to develop the element stiffness matrix based on assumed displacements [3.2]. This is equivalent to the principle of minimum potential energy that is presented in many finite element texts, but the virtual work principle is somewhat easier to understand.

The virtual work principle states that if a general structure that is in equilibrium with its applied forces deforms due to a set of small compatible virtual displacements, the virtual work done by the external forces is equal to the virtual strain energy of internal stresses. Application of this principle will produce the relations needed to solve for the equilibrium displacement configuration.

Applying this principle on the element level we have

$$\delta U_e = \delta W_e \tag{3.17}$$

where, δU_e is the virtual strain energy of internal stresses, and δW_e is the virtual work of external forces acting through the virtual displacements.

By using an assumed displacement function to define the displacement of every material point in the truss element, we can create expressions for these quantities. In general, an element is a very simple portion of the total structure or component, and then usually a very simple form of displacement function is adequate to represent the behavior of that element. Selecting a displacement function usually involves choosing a very low order polynomial. For the truss member with two nodes, the simplest form of suitable displacement function is a linear polynomial.

$$u = a_1 + a_2 x \qquad\qquad (3.18)$$

In the equation, u is the axial displacement of any material point on the truss member, a_1 and a_2 are constants to be determined, and x is the local coordinate position along the member.

Evaluate the constants by writing the equation at the node points using node values. Labeling the nodes as node i and node j the equations written at those two nodes become

$$\begin{aligned} u_i &= a_1 + a_2 x_i \\ u_j &= a_1 + a_2 x_j \end{aligned} \qquad\qquad (3.19)$$

where u_i and u_j are node displacements, and x_i and x_j are node coordinates.

With local node coordinates of $x_i = 0$ and $x_j = L$ the coefficients become

$$a_1 = u_i$$

$$a_2 = \frac{u_j - u_i}{L} \,. \qquad\qquad (3.20)$$

Using these evaluations, the equation relating displacement within the element and node point displacements is

$$u = \begin{bmatrix} 1 - \dfrac{x}{L} & \dfrac{x}{L} \end{bmatrix} \begin{Bmatrix} u_i \\ u_j \end{Bmatrix} = [N]\{d\} \qquad\qquad (3.21)$$

where $[N]$ are the element shape functions or interpolation functions, and $\{d\}$ are the node displacements.

The strain in the truss member is given by the strain displacement relation in equation (3.22). $[B]$ is an element matrix relating strain to node point displacements and consists of derivatives of the element shape functions.

$$\epsilon = \frac{du}{dx} = \frac{d[N]}{dx}\{d\} = [B]\{d\} \qquad\qquad (3.22)$$

The stress in the member follows from the stress-strain relation

$$\sigma = E\epsilon = E[B]\{d\} \tag{3.23}$$

where E is the modulus of elasticity. From these expressions we can see that the stress and strain are constant in the member. That results from assuming that the displacement of points along the member are a linear function of position, x. Also, both stress and strain are now expressed as functions of the node displacements.

Now for any given set of small virtual displacements $\{\delta d\}$ define the internal virtual strain energy δU_e by

$$\delta U_e = \int_V (\delta\epsilon)\sigma dV . \tag{3.24}$$

In this equation, $\delta\epsilon$ is the virtual strain produced by the small virtual displacements, σ is the stress level at equilibrium, and dV indicates the differential volume element of the member.

The external virtual work of nodal forces is

$$\delta W_e = \{\delta d\}^T \{f\} \tag{3.25}$$

where, $\{f\}$ are the nodal forces. Then using the principle of virtual work

$$\int_V (\delta\epsilon)\sigma dV = \{\delta d\}^T \{f\} . \tag{3.26}$$

Making the substitutions from the relations above produces equation (3.27) and by further rearranging equation (3.28).

$$\int_V [B]\{\delta d\}E[B]\{d\}dV = \{\delta d\}^T \{f\} \tag{3.27}$$

$$\{\delta d\}^T \int_V [B]^T E[B]\{d\}dV = \{\delta d\}^T \{f\} \tag{3.28}$$

Canceling $\{\delta d\}^T$ from both sides of the equation yields the element equation (3.29), where the element stiffness matrix is in equation (3.30).

$$[k]\{d\} = \{f\} \tag{3.29}$$

$$[k] = \int_V [B]^T E[B] \, dV \qquad (3.30)$$

The element stiffness matrix for the truss element becomes

$$[k] = \frac{EA}{L} \begin{bmatrix} 1 & -1 \\ -1 & 1 \end{bmatrix} \qquad (3.31)$$

because none of the matrices vary over the member volume.

This illustrates the development of the element formulation by the principle of virtual work. This one-dimensional formulation also transforms to two- or three-dimensional space via the method illustrated in the previous section. While this kind of development is not necessary for the truss element, it will be necessary for development of more complex two-dimensional and three-dimensional continuum elements in later chapters.

3.3   The Finite Element Model

The practitioner of finite element analysis normally uses an existing computer code. Because of the general complexity and sizable effort required to create a finite element code, it is impractical to consider writing a code for every specific problem that needs solving. So the job of the practitioner is to use an existing code to solve the problem of interest. The user in this case is responsible for creating the model of the structure, for managing the execution of the program, and for interpreting results created by the program.

In following the analysis procedure we reach the point of planning the model. The arrangement of nodes and elements that describe the model is known as the mesh. Using all the information that is known about the problem, and knowing the capabilities of the chosen program to analyze the problem, the mesh is planned to model the structure properly. In truss structures, each member is modeled as one truss element with the connections of truss members or elements at the node points. Based on the formulation and general assumptions for truss members these node connections behave as pinned joints. This results in no flexural loading of the member in that one member can swivel or hinge relative to the others connected at any given node, yet will transmit axial load.

Since a truss element behaves exactly in agreement with the assumptions of a truss member, there is no need to divide a member into more than one element. In fact, such a subdivision will cause the execution stage of the program to fail. The failure is due to the zero stiffness against any lateral force applied at a node connection where two members are in perfect axial alignment. So just as a physical truss structure constructed in this manner would collapse, the numerical solution of the problem defined in this manner also should collapse.

A simple bridge structure is in Figure 3-3 and a corresponding finite element mesh with elements and nodes numbered is in Figure 3-4.

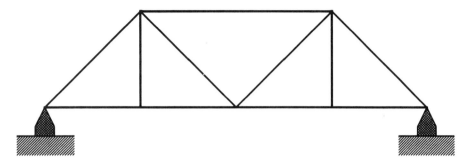

Figure 3-3. Simple Bridge Structure

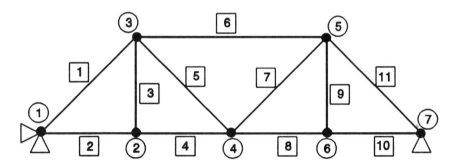

Figure 3-4. Simple Bridge Structure Finite Element Mesh

In this model there exists some geometrical symmetry. One of the ways to reduce the size of the analysis problem and computer solution is to use symmetry to reduce the model size [3.3]. Keep in mind however that loading symmetry also must exist to use symmetry conditions. If the structure in Figure 3-3 has three loads applied symmetrically as shown in Figure 3-5, then the symmetric model illustrated in Figure 3-6 will solve the problem.

To use the partial model we must impose conditions that make the partial model behave exactly as it would if it were a section of the whole

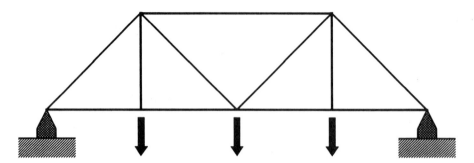

Figure 3-5. Simple Bridge Structure with Loads

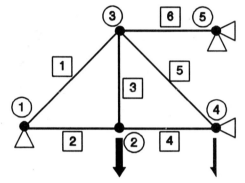

Figure 3-6. Simple Bridge Structure Finite Element Mesh

model. This usually involves applying displacement constraints and changing some of the loads to produce a solution identical to that found using the whole model. In this example, the displacement conditions to apply are that the nodes lying on the line of symmetry are constrained against displacement normal to the line of symmetry. In the actual case they would have no displacement component off the line of symmetry. The displacement components along the line of symmetry remain free to move. In this case also all loads on nodes lying on the line of symmetry are halved because there is a symmetrical portion of the model that carries a symmetrical portion of the load. In the example, the load at the middle node reduces by a half before application to the section model. This completes the definition of the model for analysis. The user must then prepare and execute the preprocessing program or, if a preprocessor is unavailable, must create an input file in the format defined by the processor program's user's guide.

3.4   Computer Input Assistance

Most programs have a preprocessor program to assist the user in the input of a model. In general, the required inputs are the data necessary to locate all the node points in their spatial configuration, a definition of the elements by the node point numbers which they connect, a definition of the type of analysis to be done, the material properties, and the displacement conditions and applied loads.

An interactive preprocessor is much preferable at this point since each node point appears directly on the computer terminal screen upon data entry. Also, each element is drawn upon definition as well as symbols for displacement conditions and loads following their data entry. Whether interactive or not, the user should prepare and assemble the data required for the model and begin execution of the input. A preprocessor should have some abilities to generate portions of the data for the mesh either through simple replications or interpolations to reduce the user's data entry. In truss models, in all likelihood, the nodes are defined by manual entry of all the nodes or of all the nodes in a repeating segment of the structure.

Most programs provide the ability to replicate nodes as a group by simple incrementation of the number set with increments of coordinate values. If a mesh calls for a line of nodes at regular spacing, then most programs allow simple interpolation along a line connecting two end nodes with the number of nodes to interpolate.

Truss element definition consists of a list of two node numbers that connect to form the element. Computer generation of additional elements can occur when there is a fixed increment between the node numbers defining one element and the node numbers defining the next consecutive elements.

The additional information needed for the element definition of a truss is its modulus of elasticity and cross section area. Usually, these data consist of one or more sets of values. Each set of these values create a material table, and then they associate with the element by table number assignments to all elements. The length of the element required to determine the element stiffness is calculated directly from the element's node coordinates.

Select nodes by number or visual position for application of displacement conditions, and specify which of its components to fix. It may have a nonzero specified value if the structure is being displacement loaded. Most fixed displacement conditions are zero. This occurs at rigid boundary supports or for enforcement of symmetry conditions. Be careful not to overconstrain the structure by prescribing zero displacements where there is no physical support.

Select nodes for application of load conditions. Load value and component direction entries associate with the selected nodes. Once the user has completed these steps, careful model checking should verify agreement with the planned model through graphic display of the model with its applied boundary conditions and loads. At this point, depending on the requirements of the specific preprocessor, the user may write input data files for the processor into computer disk files.

3.5   The Analysis Step

Operation of the analysis step is mostly transparent to the user. However, there are some factors that the user must be familiar with to assure good results. Some of them are inherent in the computer hardware and software, but some of them are under user control. For small truss structure models, most programs have enough numerical accuracy and performance to provide an accurate solution without much user concern. Large models cause the most concern [3.1].

A large model is one in which there are many elements and nodes used to represent the structure. (The structure itself is not necessarily large.) With many system equations it becomes difficult to find a numerical solution if the equation matrix is full. Even the inverse of a 10 by 10 matrix may be inaccurate if done by Gauss elimination with only a few significant figures carried along in the mathematical operations. The accuracy will improve, however, if the matrix has its nonzero terms clustered near the diagonal. This reduces the number of operations and reduces the roundoff error carried along in each operation. This kind of matrix, with its nonzero terms near the diagonal, is a banded matrix.

In a finite element model the node or element numbering pattern can strongly influence the bandwidth. In fact, a well-planned numbering pattern can minimize the matrix bandwidth. There are two general methods for setting up the order of the equations in assembly of the structure stiffness matrix [3.4]. One uses the node number numerical sequence with the degrees-of-freedom (DOF) per node to order the equations. We define the bandwidth as the measure of the closeness of terms to the diagonal in this method. The other method uses the element number numerical sequence with the node numbers of each element and the DOF per node to order the equations. We define the wavefront as the measure of closeness to the diagonal in this method. The way to minimize either of these is to plan the numbering pattern so nodes that connect through elements have their equations assembled close together in the structure stiffness matrix. This means that the node numbers used to define an element be close together numerically.

We may illustrate using the simple spring models of the following figures. First, for equations assembled by the node numerical sequence, refer to the two numbering plans in Figure 3-7. Encircled node numbers show on the models, and X's fill in the nonzero terms of the stiffness matrix. The plan on the left results in a small bandwidth, and the plan on the right results in a large bandwidth.

NODE SEQUENCED EQUATIONS

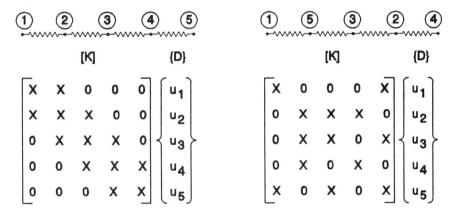

Figure 3-7. Bandwidth of Simple Stiffness Matrices

For equations assembled in the element numerical sequence, refer to the two plans in Figure 3-8. Enclosed squares denote element numbers in the models. The plan on the left results in a small wavefront, and the plan on

ELEMENT SEQUENCED EQUATIONS

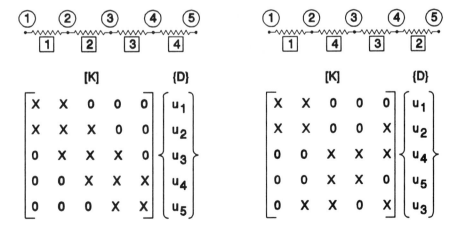

Figure 3-8. Wavefront of Simple Stiffness Matrices

the right results in a large wavefront. Note the equation order generated by the element definition sequence in the right side model.

If the truss structure model consists of thousands of nodes and elements, then the bandwidth (or wavefront) of the structure equations needs to be small. Keeping it small reduces error and computing time. If the mesh plan does not have a small bandwidth for the system of equations, then bandwidth (or wavefront) minimizers available in many programs should reduce it. There are several algorithms available which will usually, but not always, find a better node or element numbering pattern. Most programs keep the original numbering in the model for documentation and presentation purposes by storing the renumbered nodes and elements in translation tables. In some programs the user has the option to keep the original numbering or change to the new numbering plan.

Other factors that may affect the numerical performance include the mathematical and numerical algorithm quality, the numerical range of the computer, and the relative stiffness of members in the structure itself. The user has little control over the first two parts in commercially available programs except by the choice of which program to use. However, the user should be aware of the sigificance of these parts in evaluating the final results of any analysis. The algorithms for equation solution mostly derive from Gauss elimination and are subject to roundoff errors in the computer. That is why it is important to reduce the bandwidth or wavefront.

The user does have some control on the relative stiffness of adjacent members. The difficulty occurs whenever members of extremely high and low stiffness connect. This can cause the computer precision to stretch beyond its limits [3.1]. The normal precision of the computer in single-precision is about 6 significant digits, while in double precision most carry about 14 significant digits. Most finite element programs use double precision. Some computers carry greater precision than 6 or 14 digits and some have quadruple precision features, but these are not commonly used.

So if two connecting elements have stiffnesses 6 or 7 orders of magnitude different, then the precision of the terms of the system of equations becomes suspect. In an actual truss structure if two members like this existed it would be an undesirable feature of most designs. So there is likely an error in judgment in the modeling process if such a condition occurred in a truss structure model. However, this problem can occur in models using other element types where element subdivision must adequately approximate the variations occuring in the variables of solution.

The approximation error for the truss element is zero since the element formulation is in exact agreement with the assumptions used to define a truss member. During processor execution there are usually some prompts of progress made displayed on the computer and if errors occur messages appear. Sometimes these messages have meaning only to the computer program developer, but some of them can be very helpful in determining the model error.

The most common runtime errors involve incorrect definition of elements or incorrect application of displacement boundary conditions. For example, both conditions can produce an error message that the structure stiffness matrix is not positive-definite or that a negative pivot or diagonal term in the stiffness matrix appeared during equation reduction. For truss models this can occur whenever there are not enough boundary conditions to prevent rigid body motion. It can also occur when two elements connect in-line resulting in zero lateral stiffness. It can also mean that the truss structure itself is not kinematically stable associated with a kinematic linkage of the members.

The error messages from the computer program should provide an associated element number, node number, or equation number where the error occurred. That makes it easier for the user to pinpoint the problem. When execution completes without errors, then postprocessing may begin.

3.6   Output Processing and Evaluation

At this stage of the analysis, all programs have numerical results in the form of a listing file of the problem. This file will include a summary of the input data followed by numerical values of all node displacement components and all element stress results. One of the important steps to take here is to review the summary of the input printout, scanning it for errors in input interpretation of the data entered or selection of default parameters that are not appropriate for the problem at hand. This can usually be done effectively if the program formats the data for easy viewing.

The results for displacements and stresses in the listing file for large models are so lengthy that scanning is not practical. However, many programs will print a summary of maximum values for displacement components and stress magnitudes. Therefore, it is very desirable to present the data graphically for more effective evaluation.

The first graphic of importance should be an exaggerated deformed shape of the structure. All postprocessing programs will include this graphic which uses the node displacements with a scale factor to exaggerate the deformation and make it more apparent to the eye. The deflections in most engineering structures are usually very small, and without an exaggeration scale factor the deformed shape would look the same as the undeformed shape. Program options usually exist to either provide both an undeformed and deformed mesh simultaneously or an outline of the undeformed object superimposed upon the graphic of the deformed mesh.

The engineer must look at this plot critically and make sure that the boundary conditions are correct and that the shape of the deformed structure agrees qualitatively with the expected deformation. In truss structures the deformed shape will obviously show each member or element as a straight line connecting the nodes in an exaggerated deformed position.

After thorough evaluation of the deformed shape, the graphics should then turn to plots of the stress components. In continuum structures the stress component plots relate to averaged quantities at the node points. Truss structures have a stress in each member that is constant and most commercial postprocessing programs do not provide much in the way of graphic presentation of these stresses. In this event the user must return to the listing file and examine it for the highest stressed members.

The evaluation of the results determine whether we need to make additional model refinements and whether the results have converged to enough accuracy. For truss structures we know that the element formulation is exact, therefore in a correctly defined and processed model the output results will be exact. So there is no need for refined modeling to produce converged results in this case, but the modeling of loads and boundary conditions may not be fully appropriate.

It is important to remember that this is a linear elastic analysis. One of the potential failure modes is overstressing while another is elastic buckling. The stresses compare to yield strength for the material to determine if overstress failure occurs. To determine whether there is potential for elastic buckling the user must identify the members with a significant compression load, and then use Euler buckling equations from mechanics of materials to evaluate the potential for each member to buckle [3.5]. This is not done in a linear elastic analysis computer program. If any member has an inadequate safety factor against buckling, then the entire structure should have a stability analysis conducted using a solution algorithm available in some nonlinear computer codes.

3.7   *Case Studies*

We show a typical truss structure in Figure 3-9. This structure has members of two different cross section areas and two different materials. The mesh plan for this structure is illustrated in Figure 3-10 with a chosen pattern for node and element numbers. The enforced boundary conditions symbolized by the triangular shapes have the triangle tip placed on the node pointing in the direction of restraint. The loads symbolized by long arrows apply to the indicated node points.

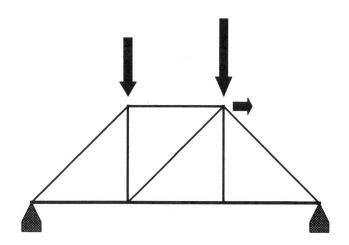

Figure 3-9. Two-Dimensional Truss Structure

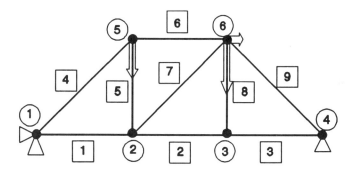

Figure 3-10. Truss Structure Finite Element Model

Table 3-1 lists the input data required to create this model. The title line is first followed by the control data line. Node definition lines begin with its number with boundary condition restraints and coordinate locations following. In this model all the z boundary conditions are restrained and all the z coordinates are zero because this is a 2-D truss element without any z degrees-of-freedom. Additional control data is next followed by load data lines. The load data lines begin with the node number of application with a direction and magnitude. The type of element is a truss. The material data is shown with two table entries. Material 1 has a modulus of elasticity for steel with a cross section area of 0.4 in², and material 2 is aluminum with a cross section area of 0.7 in². The element definitions are given by the entry of two node numbers at the end points of the element with a material table assignment.

Table 3-1. Input Data File for Truss Structure Model

```
***********¦notes inside vertical bars are for data explanation only¦
Truss Case Study                                        ¦title line¦
    6    1    1                  ¦6 nodes, 1 element group, 1 load case¦
    1    1  1  1      0.000    0.000    0.000    ¦node number,        ¦
    2    0  0  1     10.000    0.000    0.000    ¦x,y,z boundary      ¦
    3    0  0  1     20.000    0.000    0.000    ¦conditions,         ¦
    4    0  1  1     30.000    0.000    0.000    ¦0-free,1-fixed      ¦
    5    0  0  1     10.000   10.000    0.000    ¦x,y,z               ¦
    6    0  0  1     20.000   10.000    0.000    ¦coordinates         ¦
    0                            ¦number of inclined boundary conditions¦
    1    3                              ¦load case 1, 3 loads¦
    5    2  -5000.0             ¦node 5, y dir., -5000 value¦
    6    1   2000.0             ¦node 6, x dir.,  2000 value¦
    6    2  -7000.0             ¦node 6, y dir., -7000 value¦
    1    9    2        ¦element type 1-truss, 9 elements, 2 materials¦
    1  30.0E+06  0.40           ¦material 1, E = 30E6, A = 0.4¦
    2  10.0E+06  0.70           ¦material 2, E = 10E6, A = 0.7¦
    1    1    2    1    ¦element #, node I, node J, material #¦
    2    2    3    1
    3    3    4    1
    4    1    5    2
    5    5    2    2
    6    5    6    2
    7    6    2    2
    8    3    6    2
    9    6    4    2
```

This is such a small model that the equation bandwidth factor is insignificant along with any question of numerical performance or precision. Following program execution the deformed shape of the structure is shown in Figure 3-11. Note particularly that the enforced

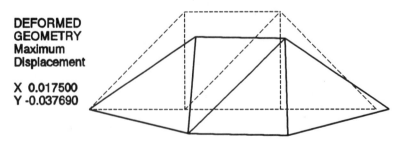

DEFORMED
GEOMETRY
Maximum
Displacement

X 0.017500
Y -0.037690

Figure 3-11. Truss Structure Deformed Shape

boundary conditions match, and the structure deforms in a manner which agrees with its expected deformation. The displacement, load and stress results are in Table 3-2. Finally, the stress results are displayed in bar graph form in Figure 3-12.

Table 3-2. Results Data File for Truss Structure Model

D I S P L A C E M E N T S

NODE	X-DISP	Y-DISP	Z-DISP
1	0.000000	0.000000	0.000000
2	0.005833	-0.035436	0.000000
3	0.011667	-0.037694	0.000000
4	0.017500	0.000000	0.000000
5	0.015233	-0.035436	0.000000
6	0.008091	-0.037694	0.000000

S T R E S S E S   I N   T R U S S   E L E M E N T
G R O U P   1

ELEM #	FORCE	STRESS
1	7000.	17500.
2	7000.	17500.
3	7000.	17500.
4	-7071.	-10102.
5	0.	0.
6	-5000.	-7143.
7	0.	0.
8	0.	0.
9	-9899.	-14142.

We may evaluate these results for the design by knowing the additional property given by its yield strength. If we say the steel has a yield strength of 50 kpsi and the aluminum has a yield strength of 30 kpsi, then the lowest factor of safety will be in element number 9 which has a value of 2.12. We see that members 4, 6, and 9 have compressive axial loads. If we use the Euler buckling equation we have to know more than just its elastic modulus and cross section area. We also must know its cross section shape to find the area moment of inertia and the end conditions. If we assume pinned end conditions and the cross section shape is a solid circular rod then the buckling loads for these elements would be 19250, 38500, and 19250 respectively. Thus, we have a reasonable factor of safety against buckling also.

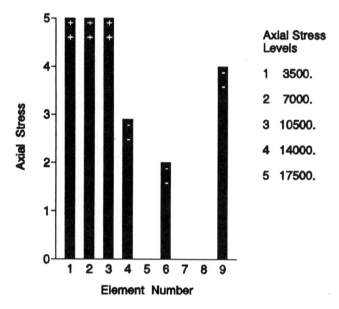

Figure 3-12. Truss Structure Member Stresses

3.8   Closure

There are few situations in mechanical design where a truss element is the right element for modeling the behavior. It is a simple element with which to discuss and learn finite element concepts. It may make an important contribution to the analysis by use as a boundary supporting spring or a gap element that connects two or more separate parts of a machine that must interact in the analysis. Therefore, the engineer must understand its nature well to interpret its effect on the overall response of any analysis that includes truss or truss-based elements.

Problems

3.1   Do the matrix multiplications shown.

$$\begin{bmatrix} 1 & -1 & 0 \\ 0 & 2 & 3 \\ 4 & 0 & -1 \end{bmatrix} \begin{bmatrix} 2 & 0 \\ 0 & 3 \\ 1 & -1 \end{bmatrix} = \qquad \begin{bmatrix} 1 & -1 & 0 \\ 0 & 2 & 3 \\ 4 & 0 & -1 \end{bmatrix} \begin{Bmatrix} 1 \\ 0 \\ 2 \end{Bmatrix} =$$

3.2   Find the solution of the following equations by using Gauss elimination.

$$\begin{bmatrix} 1 & -1 & 2 \\ 3 & 1 & 1 \\ -1 & 3 & 4 \end{bmatrix} \begin{Bmatrix} x_1 \\ x_2 \\ x_3 \end{Bmatrix} = \begin{Bmatrix} 2 \\ 6 \\ 4 \end{Bmatrix} \qquad \begin{bmatrix} 3 & 2 & 1 \\ 2 & 3 & 1 \\ 6 & 2 & 4 \end{bmatrix} \begin{Bmatrix} x_1 \\ x_2 \\ x_3 \end{Bmatrix} = \begin{Bmatrix} 3 \\ 0 \\ 6 \end{Bmatrix}$$

3.3   Perform the matrix transformation shown in eq.(P3.3) for s = sin(30) and c = cos(30).

$$[k'] = [T]^T [k][T] \qquad\qquad (P3.3)$$

where,

$$[k] = \begin{bmatrix} k & 0 & -k & 0 \\ 0 & 0 & 0 & 0 \\ -k & 0 & k & 0 \\ 0 & 0 & 0 & 0 \end{bmatrix} \qquad [T] = \begin{bmatrix} c & s & 0 & 0 \\ -s & c & 0 & 0 \\ 0 & 0 & c & s \\ 0 & 0 & -s & c \end{bmatrix}$$

3.4   A three-member truss structure is shown in Figure P3-4 with corresponding node and element numbering for a finite element model. Elements 1 and 2 are aluminum, and element 3 is steel. The cross section areas are 1.5 sq. in. for element 1 and 1.0 sq. in. for elements 2 and 3. Determine the displacement of node 2 and the stresses in each member. Solve by use of a computer program and by hand calculation. Report in a neat and concise informal engineering communication. Please include the hand calculated structure stiffness matrix in its full (8 by 8) and reduced form.

3.5   Design the derrick structure shown for a load capacity of 20 kips. Choose a suitable steel and using a factor of safety of 4.0, determine the cross section area for all the members. Recommend a cross section shape that will prevent any member from buckling.

3.6   Design a cantilevered boom to support the loads shown in the figure. All members are steel with a cross section area of 1 sq. in. The material has an allowable stress of 20 kpsi. First determine if the design is satisfactory as illustrated. Next redesign the structure within the geometric boundaries shown and the same allowable stress. A redesign may change the member arrangements, eliminate members, or change cross section areas. One of the re-design goals

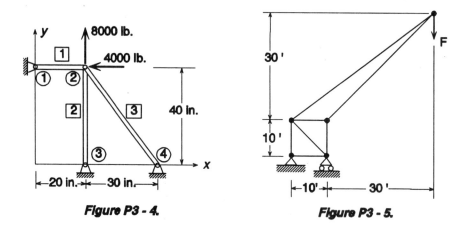

Figure P3 - 4.   **Figure P3 - 5.**

should be to reduce the overall weight of the structure. Determine a suitable cross section shape to prevent buckling for each member in compression.

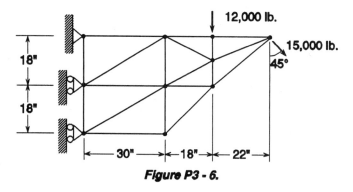

Figure P3 - 6.

3.7   Repeat Problem 3.6 for the boom in Figure P3-7.

3.8   The boxcar lift has a design capacity of 40 kips. The L-shaped member connected at D and E provides a required clearance. All members have a cross section area of 4 sq. in. and are UNS G10100 hot rolled steel. Occasional overloads are causing the L-shaped member to yield and fail. Separate analysis of the L-shaped member shows that the member can carry an 80 kip load along the line from D to E. What total structure load will produce an 80 kip load on the L-shaped member? What redesign would you propose that could provide an increase of at least 20 percent in the current factor of safety? Notes: The L-shaped member is not connected to member BE at its mid point. The axial stiffness of the L-shaped

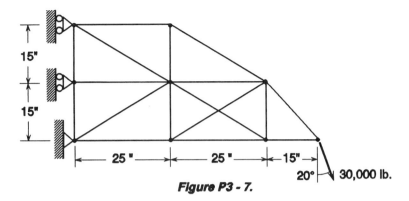

Figure P3 - 7.

member along the line from D to E is about 20 percent of the stiffness of a straight member with a 4 sq. in. cross section.

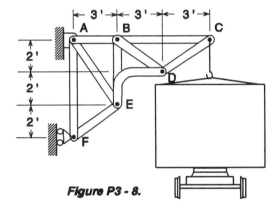

Figure P3 - 8.

3.9 Select the structural steel section(s) which will provide a safe design of the structure shown in the figure. Try to reduce weight without using an excessive number of different sections. Base the initial sizing on a truss analysis of the configuration shown. You may wish to check the stresses using a beam element model after study of the next chapter. Suggest any design changes that could improve the design or further reduce the weight.

3.10 The truss bridge in the figure is constructed of six 18 ft. partitions. The loading shown is the most severe condition of service. Determine the best steel section(s) to use for the members. Use both stress and buckling criteria. You may wish to repeat the analysis with a beam element model after study of the next chapter.

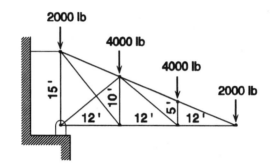

Figure P3 - 9.

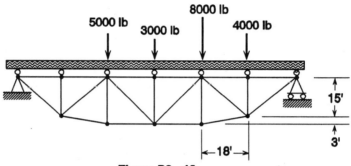

Figure P3 - 10.

References

3.1   Cook, R. D., Malkus, D. S., and Plesha, M. E., *Concepts and Applications of Finite Element Analysis*, Third Edition, John Wiley and Sons, New York, 1989.

3.2   Fenner, D. N., *Engineering Stress Analysis: A Finite Element Approach with Fortran 77 Software*, John Wiley and Sons, New York, 1987.

3.3   Akin, J. E., *Computer-Assisted Mechanical Design*, Prentice-Hall, Englewood Cliffs, New Jersey, 1990.

3.4   Bathe, K. J., *Finite Element Procedures in Engineering Analysis*, Prentice-Hall, Englewood Cliffs, New Jersey, 1981.

3.5   Popov, E. P., *Introduction to Mechanics of Solids*, Prentice-Hall, Inc., Englewood Cliffs, New Jersey, 1968.

C H A P T E R 4

BEAMS AND FRAMES

Application of straight beam theory readily solves simple beam problems especially if the problem is statically determinant. If the beam is not particularly simple, in that it may have cross section changes, multiple supports, or complex loading distributions, then we can use beam theory, but it is very tedious to develop the solution by hand. Also, many 2-D or 3-D framework structures may require solutions in which the truss member assumption is inadequate and therefore needs the beam flexure formulation. Further applications may include beam members as reinforcement members in combination beam, plate, and shell structures. These applications are readily attacked with the finite element formulation.

4.1 Element Formulation

The governing differential equation for any beam span without distributed loading has for a solution a cubic polynomial in x, where x is the coordinate position along the beam length. The solution is for the lateral displacement as a function of x. The solution equation is

$$v = a_1 + a_2 x + a_3 x^2 + a_4 x^3 .$$ (4.1)

There are four constants to evaluate by node values. For a two-node element the node values will be the lateral displacement and slope or rotation angle at each node. By expression of the displacement equation above and its derivative equation at the node locations, we can rearrange it to give the displacement as a function of x and the node values. Proceeding from this development can lead to the element formulation through the principle of virtual work.

However, use of the virtual work approach for this element gains little, and it is perhaps more physically understandable to follow the direct approach for formulating the element stiffness matrix [4.1]. The element equations relating general displacement and force components are given by

$$[k]\{d\} = \{f\} \tag{4.2}$$

where $[k]$ is the element stiffness matrix, $\{d\}$ is the node displacement component column matrix and $\{f\}$ is the internal force component column matrix. The stiffness matrix terms derive from superposition of simple beam solutions. Apply a unit displacement of one component with the other components held to zero and evaluate the magnitude of resulting force components. For example, taking the element shown in Figure 4-1 and applying a unit vertical displacement v_i with $\phi_i = v_j = \phi_j = 0$ results in the deformation illustrated in Figure 4-2.

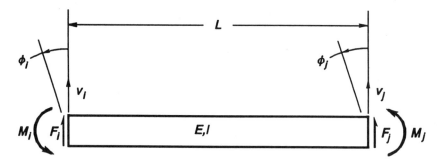

Figure 4-1. Beam Element

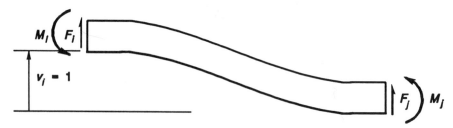

Figure 4-2. Deformed Beam Element

The forces shown which are the superposition of the solutions for a cantilever beam with an end load and an end moment produce this deformation. The superposition is done to yield a unit value of lateral displacement with a zero slope at the end. The element equations written in matrix form yield equation (4.3), which in turn yield the relations in equation (4.4).

$$
\begin{bmatrix}
k_{11} & k_{12} & k_{13} & k_{14} \\
k_{21} & k_{22} & k_{23} & k_{24} \\
k_{31} & k_{32} & k_{33} & k_{34} \\
k_{41} & k_{42} & k_{43} & k_{44}
\end{bmatrix}
\begin{Bmatrix}
1 \\
0 \\
0 \\
0
\end{Bmatrix}
=
\begin{Bmatrix}
F_i \\
M_i \\
F_j \\
M_j
\end{Bmatrix}
\tag{4.3}
$$

$$
k_{11} = F_i \ , \quad k_{21} = M_i \ , \quad k_{31} = F_j \ , \ and \ \ k_{41} = M_j
\tag{4.4}
$$

Using superposition of beam deflection equations available in any mechanics of materials text we write equations (4.5). Solve these equations for the values of F_i and M_i in equations (4.6).

$$
v_i = 1 = \frac{F_i L^3}{3EI} - \frac{M_i L^2}{2EI}
$$

$$
\phi_i = 0 = \frac{F_i L^2}{2EI} - \frac{M_i L}{EI}
\tag{4.5}
$$

$$
F_i = \frac{12EI}{L^3}
$$

$$
M_i = \frac{6EI}{L^2}
\tag{4.6}
$$

Use static equilibrium equations to get the values of F_j and M_j in equation (4.7).

$$
F_j = -\frac{12EI}{L^3} \qquad M_j = \frac{6EI}{L^2}
\tag{4.7}
$$

We now have all the terms of column 1 of the 4x4 element stiffness matrix as shown in equation (4.8).

$$[k] = \begin{bmatrix} \dfrac{12EI}{L^3} & \cdots & \cdots & \cdots \\[2ex] \dfrac{6EI}{L^2} & \cdots & \cdots & \cdots \\[2ex] -\dfrac{12EI}{L^3} & \cdots & \cdots & \cdots \\[2ex] \dfrac{6EI}{L^2} & \cdots & \cdots & \cdots \end{bmatrix} \qquad (4.8)$$

Similarly applying a unit value of rotation for ϕ_i and fixing all other components to zero we derive the force and moment values in Figure 4-3. These come from superposition of the same solutions for end load and moment to satisfy the displacement conditions.

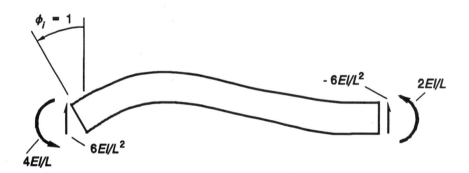

Figure 4-3 Deformed Beam Element.

Notice that the sign convention employed here is common in the finite element formulation such that the component's sign always agrees with the positive direction of a right-handed coordinate system. This does not agree with most beam sign conventions employed in mechanics of material texts. Therefore the user should be aware that the output components will normally be expressed using this finite element sign convention. This means, for example, that a positive value of moment at the first node of the element will produce a tensile stress at the top surface of the beam.

In contrast, a positive moment on the second node will produce a compressive stress at the top surface of the beam.

Obtain the remaining terms in the stiffness matrix by application of the same procedures to the second node. The final element stiffness is then given in equation (4.9).

$$[k] = \begin{bmatrix} \dfrac{12EI}{L^3} & \dfrac{6EI}{L^2} & -\dfrac{12EI}{L^3} & \dfrac{6EI}{L^2} \\ \dfrac{6EI}{L^2} & \dfrac{4EI}{L} & -\dfrac{6EI}{L^2} & \dfrac{2EI}{L} \\ -\dfrac{12EI}{L^3} & -\dfrac{6EI}{L^2} & \dfrac{12EI}{L^3} & -\dfrac{6EI}{L^2} \\ \dfrac{6EI}{L^2} & \dfrac{2EI}{L} & -\dfrac{6EI}{L^2} & \dfrac{4EI}{L} \end{bmatrix} \qquad (4.9)$$

This formulation provides an exact representation of a beam span within the assumptions involved in straight beam theory provided there are no loads applied along the span. Therefore, in modeling considerations, place a node at all locations where concentrated forces, or moments, act in creating the element assembly. In spans where there is a distributed load, the assumed displacement field does not completely satisfy the governing differential equation therefore the solution is not exact but approximate. One approach to modeling in this area is to make enough subdivisions of the span with distributed load to lessen the error. If a work equivalent load set acting on the nodes replaces the distributed load, then the influence of any error in this element will not propagate to other elements. In other words, the displacement components at the nodes will be correct if we use the equivalent load set. The equivalent load components for a distributed load on the element span are the negative of the end reaction force and moment found in the solution of a fixed end beam with the same distributed load as shown by Logan [4.2].

This formulation provides the ability to analyze simple beams, but does not account for the axial load that may exist in beam members connected in a framework. By adding the truss element formulation by superposition with the previous formulation, we have an element that can support both lateral and axial loads. The axial stiffness terms at each node are added to the element stiffness matrix formulation to create the frame element stiffness matrix in equation (4.10).

This assumes that superposition is valid for this case. If displacements are small it will be accurate, however there is an interaction that occurs between axial and lateral loading on beams. If the axial load is tensile it reduces the effect of lateral loads, and when the axial load is compressive

$$[k] = \begin{bmatrix} \dfrac{AE}{L} & 0 & 0 & -\dfrac{AE}{L} & 0 & 0 \\[3mm] 0 & \dfrac{12EI}{L^3} & \dfrac{6EI}{L^2} & 0 & -\dfrac{12EI}{L^3} & \dfrac{6EI}{L^2} \\[3mm] 0 & \dfrac{6EI}{L^2} & \dfrac{4EI}{L} & 0 & -\dfrac{6EI}{L^2} & \dfrac{2EI}{L} \\[3mm] -\dfrac{AE}{L} & 0 & 0 & \dfrac{AE}{L} & 0 & 0 \\[3mm] 0 & -\dfrac{12EI}{L^3} & -\dfrac{6EI}{L^2} & 0 & \dfrac{12EI}{L^3} & -\dfrac{6EI}{L^2} \\[3mm] 0 & \dfrac{6EI}{L^2} & \dfrac{2EI}{L} & 0 & -\dfrac{6EI}{L^2} & \dfrac{4EI}{L} \end{bmatrix} \qquad (4.10)$$

it amplifies the effect of lateral loads. To gain further information on this interaction consult an advanced mechanics of materials text [4.3] for the equations that apply to members called beam-columns or struts. The equations for these members are a nonlinear function of the size of lateral displacement. Therefore, a linear analysis cannot account for the effect.

The user should be aware of this consideration and remember that if the axial load is tensile the results from beam elements will be higher than they actually are so the results are conservative. While if the axial load is compressive the results will be less than actual and may be in serious error. The size of error associated with the compressive loading is normally quite small until the axial load exceeds roughly 25 percent of the Euler column buckling load. In most cases a design should have a factor of safety against buckling greater than four anyway.

Now the formulation includes the u and v displacement components and the section rotation at the nodes in the element local coordinate system. Using the coordinate transformations developed for truss members, we may orient this two-dimensional beam element in 2-D space. Through this transformation then the element formulation applies to any 2-D framework.

To develop a 3-D beam element, we must add the capability for torsional loads about the axis of the line element as well as flexural loads acting in the x-z plane. Set up the element in a local coordinate system with x along the line of the element, with y as one lateral direction, and z as the orthogonal lateral direction. Add the torsional response simply by superposition of the simple strength of materials torsion relationships between applied torque and the angle of twist. For the two-node element these relations are given by the following equations in matrix form

$$
\begin{bmatrix}
\dfrac{JG}{L} & -\dfrac{JG}{L} \\[3mm]
-\dfrac{JG}{L} & \dfrac{JG}{L}
\end{bmatrix}
\begin{Bmatrix}
\phi_{xi} \\[2mm]
\phi_{xj}
\end{Bmatrix}
=
\begin{Bmatrix}
T_i \\[2mm]
T_j
\end{Bmatrix}
\tag{4.11}
$$

where, J is the torsional constant about the x axis, G is the shear modulus of elasticity of the material, L is the element length, ϕ_{xi} and ϕ_{xj} are the nodal degrees-of-freedom measured by the angle of twist at each node about the local x axis, and T_i and T_j are the torques or moment about the x axis at each node.

The addition of flexure in the x-z plane adds another stiffness matrix similar to the one in equation (4.9) except that the moment of inertia is about the y axis passing through the cross section neutral axis.

The superposition of these terms leads to an element stiffness matrix of size 12 by 12 associated with the six degrees-of-freedom at each node. The matrix is given in equation (4.12).

This defines the element stiffness matrix for a 3-D beam element in a local coordinate system positioned with its origin at the first node of the element with x axis pointing to the second node. To orient the element in 3-D space, we must make coordinate transformations similar to those made for the truss element.

For a beam that is relatively long compared with its cross section dimensions the displacement is almost entirely due to the flexure of the beam section. However, for relatively short beams an additional component of the lateral displacement may be due to the transverse shear strain that exists in the beam section due to transverse shear load. Some commercial finite element programs include an approximation of this component of deformation in the element formulation. These will call for the analyst to input a shear deformation constant [4.4]. The shear deformation constant has a value associated with the geometry of the cross section. Some typical values for the shear deformation constant are 6/5 for a rectangle, 10/9 for a solid circular section and 2 for a thin-wall tube. Obviously an input value of zero will negate any shear deformation.

The limitations on this element are that we have used the same assumptions used in conventional beam and torsion theories in the element formulation. Therefore the finite element solution cannot be any more accurate than a conventional beam analysis. The axial load capability provides the ability to do framework analysis, however, the formulation does not couple the axial and lateral loading while in actual fact there is a nonlinear coupling between them. The user must also be aware that the

$$\begin{bmatrix}
\frac{AE}{L} & 0 & 0 & 0 & 0 & 0 & -\frac{AE}{L} & 0 & 0 & 0 & 0 & 0 \\[6pt]
0 & \frac{12EI_z}{L^3} & 0 & 0 & 0 & \frac{6EI_z}{L^2} & 0 & -\frac{12EI_z}{L^3} & 0 & 0 & 0 & \frac{6EI_z}{L^2} \\[6pt]
0 & 0 & \frac{12EI_y}{L^3} & 0 & -\frac{6EI_y}{L^2} & 0 & 0 & 0 & -\frac{12EI_y}{L^3} & 0 & -\frac{6EI_y}{L^2} & 0 \\[6pt]
0 & 0 & 0 & \frac{JG}{L} & 0 & 0 & 0 & 0 & 0 & -\frac{JG}{L} & 0 & 0 \\[6pt]
0 & 0 & -\frac{6EI_y}{L^2} & 0 & \frac{4EI_y}{L} & 0 & 0 & 0 & \frac{6EI_y}{L^2} & 0 & \frac{2EI_y}{L} & 0 \\[6pt]
0 & \frac{6EI_z}{L^2} & 0 & 0 & 0 & \frac{4EI_z}{L} & 0 & -\frac{6EI_z}{L^2} & 0 & 0 & 0 & \frac{2EI_z}{L} \\[6pt]
-\frac{AE}{L} & 0 & 0 & 0 & 0 & 0 & \frac{AE}{L} & 0 & 0 & 0 & 0 & 0 \\[6pt]
0 & -\frac{12EI_z}{L^3} & 0 & 0 & 0 & -\frac{6EI_z}{L^2} & 0 & \frac{12EI_z}{L^3} & 0 & 0 & 0 & -\frac{6EI_z}{L^2} \\[6pt]
0 & 0 & -\frac{12EI_y}{L^3} & 0 & \frac{6EI_y}{L^2} & 0 & 0 & 0 & \frac{12EI_y}{L^3} & 0 & \frac{6EI_y}{L^2} & 0 \\[6pt]
0 & 0 & 0 & -\frac{JG}{L} & 0 & 0 & 0 & 0 & 0 & \frac{JG}{L} & 0 & 0 \\[6pt]
0 & 0 & -\frac{6EI_y}{L^2} & 0 & \frac{2EI_y}{L} & 0 & 0 & 0 & \frac{6EI_y}{L^2} & 0 & \frac{4EI_y}{L} & 0 \\[6pt]
0 & \frac{6EI_z}{L^2} & 0 & 0 & 0 & \frac{2EI_z}{L} & 0 & -\frac{6EI_z}{L^2} & 0 & 0 & 0 & \frac{4EI_z}{L}
\end{bmatrix}$$

$$(4.12)$$

analysis does not account for stress concentration factors at cross section changes nor where point loads are applied, nor where beam frame components are connected.

4.2 The Finite Element Model

In planning the mesh for a structure to be modeled with beam elements, the factors just revealed in element formulation provide guidance about the proper element subdivision and connections. Since the element formulation is exact for a beam span with no intermediate loads, then we need only one element to model any member of the structure that has constant cross section properties and no intermediate loads. Where a span has a distributed load, we may subdivide it with several elements to lessen the error depending upon the solution accuracy desired.

There should be a node placed at every location in the structure where a point load is applied. Also, where frame members connect such that the line element changes direction or cross section properties change we should place a node and end an element at that point. Remember that the connection of two or more elements at a node guarantees that each element connecting at that node will have the same value of linear and rotation displacement components at that node. Physically, think of this as a solid, continuous or welded configuration.

From the theory of straight beams [4.3] we know that the moment at any position along the beam relates to the second derivative of the elastic or lateral displacement curve. Therefore, in the element formulation we can see that the moment distribution along any element is linear. Then, the maximum value of moment in any element must occur at one of the end nodes. This is further justification for not subdividing any unloaded span. In summary, construct the mesh for a beam or frame structure by placing nodes at the points where point load is applied, where different members connect, and within spans where distributed loads are applied.

The boundary conditions for beam element models may consist of restraints on the linear displacement components and the rotations. Simple supports would only restrain the linear displacement components while built-in supports would also restrain rotation components. For example, on the simply supported beam with an intermediate load in Figure 4-4 with its finite element model in Figure 4-5, we would fix the vertical and horizontal displacement components at nodes 1 and 3, but leave the rotation components at nodes 1 and 3 free.

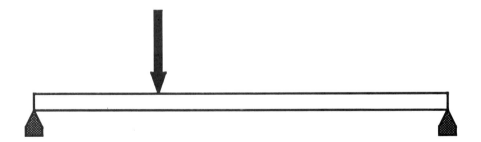

Figure 4-4. Simply Supported Beam

The cantilever beam modeled in Figure 4-6 would have both x and y linear displacement components and the rotation at node 1 fixed. Note that in the case of the simple beam if we only fixed the vertical displacement components leaving the horizontal components free, the structure

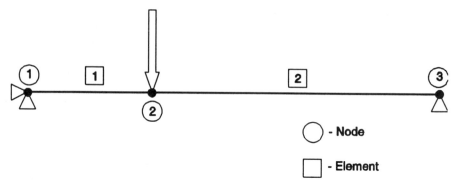

Figure 4-5. Simply Supported Beam Finite Element Model

would be free to move as a rigid body in the x direction. This would allow rigid body motion, and the program would be likely to fail.

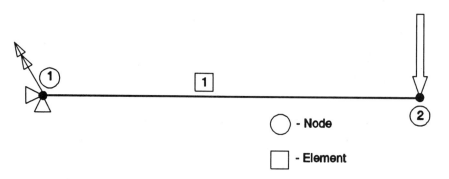

Figure 4-6. Cantilever Beam Finite Element Model

Point loads or moments may be applied at any node location. However, in real applications there is no point load or moment, but only distributed loads applied over very small areas. In this case, the structure outside the immediate vicinity of the load application does behave as if it were loaded by point load.

We mentioned in the previous section that the element formulation assumes that there are no loads along the span of the element, therefore a distributed load violates that assumption. However, we may produce an exact solution for the remainder of the structure by applying the equivalent load set for the distributed load. Some examples of the nodal load sets to replace the distributed loading are shown in Figure 4-7. Additional cases are given by Logan [4.2].

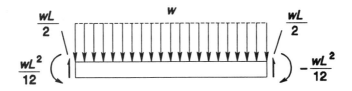

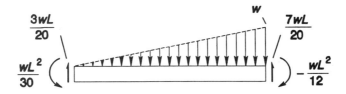

Figure 4-7. Distributed Loading Equivalent Load Sets

4.3   Computer Input Assistance

A program preprocesser will usually have the same capabilities for model input of beam structures as exists for trusses. Nodes are input by geometric location, and typically the only generation capability is either for a series of nodes along a line or replication of a group of defined nodes with geometric increments. Beam element definition consists of the two node numbers which it connects and its associated material and physical properties. The first material property of interest is its modulus of elasticity, and if a dynamic or thermal analysis is to be done its mass density and thermal expansion coefficient are also needed. The physical properties include the cross sectional area, two area moments of inertia, a torsion constant, and the location of stress calculation points.

Boundary condition specification involves simply specifying the node numbers and displacement components that have restraints. Loads are applied also by identifying node numbers and specifying the load components and magnitudes. Most commercial programs provide the ability to input a specified distributed load on a beam element, but then the equivalent load set described in Figure 4-7 will replace it within the program. When the model input is complete check the graphic display carefully for the complete definition of nodes, elements, boundary conditions, and loads.

4.4   *The Analysis Step*

If the model is small or you plan the node and element numbering well, then there should be no need for bandwidth or wavefront minimization. There is some potential for ill-conditioning of the structure stiffness matrix in beam models due to the axial stiffness term normally being much higher than the flexural stiffness terms [4.1]. This is especially severe for long slender beam elements. If ill-conditioning produces some numerical error in the results, more element subdivision in long spans will reduce it. Connecting a very short element to a very long element or large changes in moment of inertia or material stiffness from one element to the next also causes ill-conditioning.

4.5   *Output Processing and Evaluation*

A complete printout, or listing file, lists a reflection of model input data, the displacement results including rotations, and output of stresses resulting from moment, axial, and shear forces. The graphical presentation of the deformed shape ideally would use the rotations at the nodes with the assumed displacement shape function for the element to plot the actual curved shape the elements take when loaded. However, most programs only plot the deformed shape using the node translation displacements and straight line connections to represent the elements. In this case it is difficult to determine from the graphic if we applied the rotational boundary conditions. In order to check boundary conditions and get a smooth visualization of the deformation curvatures, the user may resort to remodeling with several element subdivisions within each span.

The stresses in 2-D beam elements consist of a normal stress acting normal to the beam cross section and a transverse shear stress acting on the face of the cross section. The normal stress comes from superposition of the axial stress that is uniform across the section with the bending stress due to the moment on the section. This combination will result in the maximum normal stress occurring either at the top or bottom surface. The transverse shear stress is usually an average across the cross section calculated by the transverse load divided by the area. This obviously does not account for the shear stress variation that occurs across the section from top to bottom [4.3]. The transverse shear stress must be zero at the top and bottom surfaces and has some nonuniform distribution in between which is a function of the cross section geometry. This variation is usually of minor importance, but the analyst may calculate it if desired.

When using the 3-D beam element the normal stress must be a combination of the axial stress and the flexural stress from both the local y-moment and z-moment acting on the cross section. Since the stresses due to the moments are linear functions from one edge to the opposite edge, their combination is usually highest at the extreme corners of the cross section farthest from the centroid. So depending on the sign of the moments and the cross section shape there may be several points to calculate the stress to find the maximum stress. If torsion also exists in the 3-D beam element then we must consider the shear stress that results from the torsion at the point of maximum normal stress. The combination creates a two-dimensional stress state to evaluate through use of a proper failure theory.

Most of the available finite programs do not make graphical presentation of the beam stress results. So it reverts to the engineer to evaluate the stress output usually based upon values from the printout listing. The engineer also must check for Euler buckling in members that have an axial compressive stress. If the factor of safety against buckling in these members is less than about 4, then the stresses may need correction for the interaction between the axial and flexural stress in that member.

4.6 *Case Studies*

We show a simple beam structure in Figure 4-8 with two different cross sections and two loads. It has simple supports, and we develop a mesh plan in Figure 4-9 with five nodes and four elements. The model input data list is in Table 4-1. The title line is first with the control data line next. Node definition begins with its number with boundary condition restraints and coordinate locations following. The z boundary condition for the 2-D beam element applies to the rotation degree-of-freedom. All the z boundary conditions are free in this model.

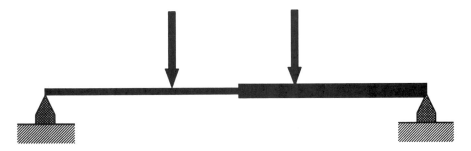

Figure 4-8. Simple Beam with a Cross Section Change and Two Loads

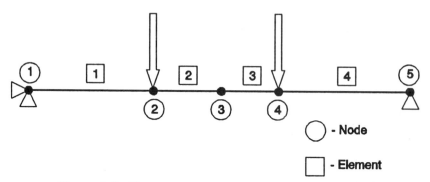

Figure 4-9. Finite Element Model of the Simple Beam

Table 4-1. Input Data File for Beam Model

```
***********|notes inside vertical bars are for data explanation only|
Beam Case Study                                            |title line|
    5    1    1                |5 nodes, 1 element group, 1 load case|
    1    1  1  0     0.000   0.000    0.000    |node number,         |
    2    0  0  0     5.000   0.000    0.000    |x,y,z boundary       |
    3    0  0  0     8.000   0.000    0.000    |conditions           |
    4    0  0  0    10.000   0.000    0.000    |0-free, 1-fixed       |
    5    0  1  0    15.000   0.000    0.000    |x,y,z coord.          |
    0                        |number of inclined boundary conditions|
    1    2                              |load case 1, 2 loads|
    2    2  -300.0           |node 2, y direction, -300 magnitude|
    4    2  -150.0           |node 4, y direction, -150 magnitude|
    2    4    2     |element type 2-beam, 4 elements, 2 materials|
    1    30.0E+06   0.40   0.04   0.50   |matl 1, E=30E6, A=0.40|
    2    10.0E+06   0.60   0.06   0.75   |       I=0.04, c=0.50|
                                    |matl 2, E=10E6, A=0.60|
                                    |       I=0.06, c=0.75|
    1    1    2    1         |element #, node i, node j, material #|
    2    2    3    1
    3    3    4    2
    4    4    5    2
```

Load specification consists of the node number of application with the load direction and magnitude. The type of element is a beam. The material data is shown with two table entries. Entry 1 has a modulus of elasticity for steel with a cross section area of $0.4 \ in^2$, moment of inertia of $0.04 \ in^4$, and distance from neutral axis to the surface of 0.5 in. Entry 2 is aluminum with a cross section area of $0.7 \ in^2$, moment of inertia of 0.06 in^4, and distance from neutral axis of 0.75 in. The element definitions are given by the entry of two node numbers at the end points of the element with a material table assignment.

Following execution of the program the deformed shape plot appears in Figure 4-10, and the results printout is in Table 4-2. In this table the z displacements are the angular rotations of the nodes. The axial stress is the value from any axial load acting on the element. The flexure stress is due to the resulting bending moment and is the value on the beam top surface when the element definition has nodes I and J arranged left to right. The average shear stress is simply the transverse shear load divided by the cross section area. The shape of the cross section determines the actual shear stress distribution across the beam height. Finally a bar graph display of the beam element stresses is given in Figure 4-11.

**DEFORMED
GEOMETRY**
Maximum
Displacement

X 0.0000
Y -0.0316

Figure 4-10. Deformed Shape of the Simple Beam

Table 4-2. Results Data File for the Beam Model

D I S P L A C E M E N T S

NODE	X-DISP	Y-DISP	Z-DISP
1	0.000000	0.000000	-0.006026
2	0.000000	-0.025789	-0.003422
3	0.000000	-0.031555	-0.000484
4	0.000000	-0.028968	0.003016
5	0.000000	0.000000	0.007182

S T R E S S E S   I N   B E A M   E L E M E N T
G R O U P   1

ELEM #	AXIAL STRESS	FLEXURE STRESS NODE I	FLEXURE STRESS NODE J	AVG SHEAR STRESS
1	0.	0.	-15625.	625.
2	0.	-15625.	-13750.	-125.
3	0.	-13750.	-12500.	-83.
4	0.	-12500.	0.	-333.

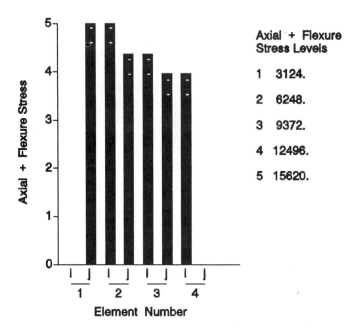

Figure 4-11. Beam Element Stress Display

These results show that while the finite element method provides solutions as valid as straight beam theory will allow, there has been no accounting for stress concentration effects where the cross section change occurred.   The designer's job here is to take the results from these analyses and then do more detailed modeling of the exact configuration where the cross section change occurred to evaluate the potential for failure at that location.

4.7   Closure

The use of beam elements in models provides the engineer with the opportunity to solve rather complex beam structures or frameworks which could not easily be done with conventional approaches. It also can include the effects of the stiffness of supporting structures through connection with truss elements or beam elements selected to approximate the support stiffness.   In the case of statically indeterminant structures where the supports might have different stiffnesses the finite element model will provide much better solutions than we can get by conventional approaches.   It also can provide the loading that exists at localized areas where cross section changes or member connections occur. Then we can use the loading in much more detailed models of those regions.

Problems

4.1 The structure shown in Figure P4-1 has a horizontal steel beam welded to a rigid column on the left and simply supported on the right end. There is also a steel rod with pinned attachments to the column and the beam providing support for the beam. The beam cross section is shown on the right, and the rod diameter is 25mm. Evaluate the effectiveness of the steel rod for reducing stress in the beam by analyzing models with and without the rod and comparing results.

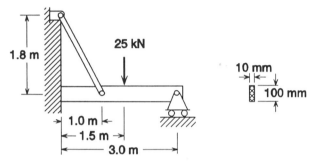

Figure P4 - 1.

4.2 Curved beams can be approximated by a group of straight beam elements. Opposed diametral forces load the thin circular ring in Figure P4-2. Use symmetry and determine the number of elements required to achieve 5 percent accuracy for the maximum moment and the displacements along and perpendicular to the loaded diameter. Assume that the thickness, t = r/10, and that the cross section of the ring is square. The maximum moment and radial displacements are given by

$$M_a = \frac{Pr}{\pi} \qquad \delta_a = \frac{Pr^3}{8EI}\left(\pi - \frac{8}{\pi}\right) \qquad \delta_b = -\frac{Pr^3}{4EI}\left(\frac{4}{\pi} - 1\right)$$

where, a is the diameter parallel to the loads, and b is perpendicular to the loads.

4.3 Analyze the bicycle frame design sketched in figure P4-3. Use a vertical load of 150 lb. at the seat location and 25 lb. at the handlebar location and apply a load factor of 2.5 for inertial loading. Assume for the first analysis that all the members are tubular steel with a 1 in. outside diameter and 0.062 in. wall thickness. From the

first analysis determine if any yield failures are likely if the material
is a high-carbon steel with a yield strength of 110 kpsi. If yielding
will occur, refine the design by replacement of highly stressed
members with a more substantial section or by altering the design
layout to eliminate yield failures. If the frame is overdesigned,
refine the design to reduce weight. Do the deflections seem
excessive? Is there a specific location that seems to be too flexible?

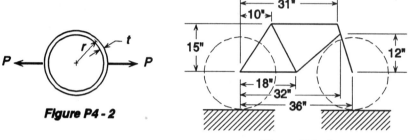

Figure P4 - 2

Figure P4 - 3.

4.4   Design the support arms, AB and CD, for the lift platform illustrated
in Figure P4-4. Select a carbon steel material and choose a suitable
cross section shape with no more than a 4 to 1 ratio of moments of
inertia between the two principal directions of the cross section.
You may allow the cross section dimensions to vary along the length
of the arms to reduce their weight. The actual structure has four
support arms, but the loads shown are for one pair of arms. The
loads values are for operating conditions and should have a proper
factor of safety applied for human safety. In developing finite
element models, remove the platform and replace it with statically
equivalent loads at the joints at B and D. Use truss elements or
beam elements with low flexural stiffness to model the arms from
B to D, the intermediate connection, and the hydraulic actuator.

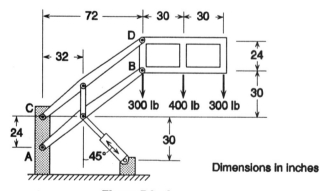

Figure P4 - 4.

4.5 There is a proposal to build the two-story framework shown in Figure P4-5. The members are all to be steel I-beams with rigid connections. The beams are 12" I 50.0 lb/ft except for the floor joists that are 18" I 70.0 lb/ft. The material is UNS G10100 hot-rolled steel. Two horizontal loads and vertical floor loads are shown. Is this a viable design? Are the deflections reasonable? Are any members overdesigned?

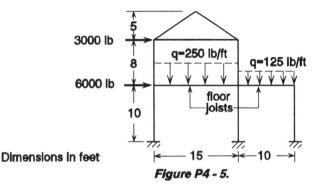

Figure P4 - 5.

4.6 Figure P4-6 shows an initial design for a pulpwood loader. The average weight for the maximum number of logs picked up by the loader is about 10 kN. Select a steel and determine a suitable tubular cross section for the main upright member, BF, that has attachments for the hydraulic cylinder actuators, AE and DG. Select a steel and find a viable hollow rectangular cross section for the horizontal load arm, AC. The rectangular section may vary in dimension along the member length to reduce weight. The finite element models may use beam elements for all members but the hydraulic cylinders which should be truss elements. The pinned joint at B between the upright and horizontal members is best modeled with a release of the end node of the top element on the upright member. However, if that capability doesn't exist in the computer program, you may have to trick it with low flexural stiffness or separate the models of the upright and horizontal members since the interface forces are statically determinant here.

4.7 Model the compression spring in Figure P4-7 using 3-D beam elements. The material is A228 music wire. It has an inside diameter of 0.8 in., a wire diameter of 0.100 in., and a free length of 3.0 in. There are 10 total coils but only 8 active coils as the ends are squared and ground. Determine the axial spring constant and compare with the common equation for calculating the spring constant. Determine the lateral spring constant for loads perpendicular to the spring coil axis at the ends. Compare the stresses from the finite element model with conventional calculations.

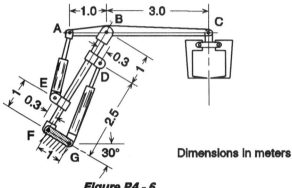

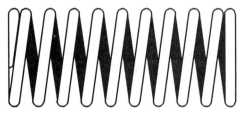

Dimensions in meters

Figure P4 - 6.

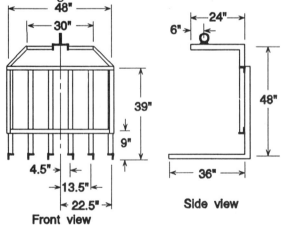

Figure P4 - 7.

4.8   A concrete block loading and unloading dolly is drawn in Figure P4-8. The top center member with the eye for the loading hook is a 6 in. structural steel channel section, 6 C 8.2 lb/ft.   All the other members are 2 x 2 x ¼-in. structural steel angle.   The material is UNS G10100 hot-rolled steel.   The maximum design load capacity is to be 2250 lb.   Is this a safe design?   Are there critical spots in the design that need to be redesigned?

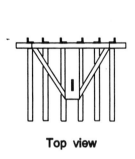

Top view

Front view
Figure P4 - 8.

References

4.1 Cook, R. D., Malkus, D. S., and Plesha, M. E., *Concepts and Applications of Finite Element Analysis*, Third Edition, John Wiley and Sons, New York, 1989.

4.2 Logan, D. L., *A First Course in the Finite Element Method*, PWS-KENT Publishing Co., Boston, Massachusetts, 1986

4.3 Cook, R. D., and Young, W. C., *Advanced Mechanics of Materials*, Macmillan Publishing Co., New York, 1985.

4.4 Popov, E. P., *Introduction to Mechanics of Solids*, Prentice-Hall, Inc. Englewood Cliffs, New Jersey, 1968.

CHAPTER 5

TWO-DIMENSIONAL SOLIDS

While the finite element method is very helpful for the solution of truss, beam, and frame problems, the real power of the method shows in application to two- and three-dimensional solid analysis. There are very few closed form solutions to two-dimensional problems, and they are only available for simple geometries and loading conditions. The finite element method on the other hand, if correctly applied, can provide the solution to most any two-dimensional problem. The correct application is of prime importance, and the analyst makes decisions involving the layout and planning of the model to represent the member under analysis. The correct application must be done to limit solution errors.

Equations derived in theory of elasticity govern the solution to problems in two-dimensions. The finite element formulation must satisfy, at least approximately, the relations between displacements, strains, and stresses to find a solution for general two-dimensional problems.

5.1   Element Formulation

We must use either the principle of virtual work introduced in Chapter 3 from the finite element application by Fenner [5.1] or the equivalent principle of minimum potential energy for element formulation in 2-D and 3-D continuum structures. In this chapter we use the virtual work principle to develop a general form of the element stiffness matrix. The general formulation leads to development of the triangle and quadrilateral two-dimensional finite elements. The key to formulation is making an assumption about the mathematical function of material displacement throughout the element volume.

The virtual work principle states that if a general structure which is in equilibrium with its applied forces is subjected to a set of small, compatible virtual displacements, the virtual work done by the external forces is equal to the virtual strain energy of internal stresses. Application of this principle will produce the relations needed to solve for the equilibrium displacement configuration.

Applying this principle on the element level we have

$$\delta U_e = \delta W_e \tag{5.1}$$

where, δU_e is the virtual strain energy of internal stresses, and δW_e is the virtual work of external forces acting through the virtual displacements.

Using an assumed displacement function for the displacement of any material point throughout the element approximately satisfies the equations of elasticity. The displacement function and node point values of displacement prescribe the displacement of every material point throughout the structure. Usually an element is a small portion of the structure for analysis, and then a very simple form of displacement function adequately represents the behavior of that element. Selecting a displacement function usually involves choosing a very low order polynomial.

Assume that the field displacement components at any material point on the element field interpolate from the node point values. The field displacement components equations are then

$$\{u\} = [N]\{d\} \tag{5.2}$$

where $\{u\}$ are the field displacement components (which are u and v in 2-D and u, v, and w in 3-D), $[N]$ are the interpolation functions, and $\{d\}$ are the node point displacement component values for the element.

The interpolation functions are chosen based on element geometry and the assumed form of the displacement function within the element. These will be shown in specific element formulations.

The strain displacement relations that were given in Chapter 2 for 2-D components define the strain at any material point. Application of these relations to the assumed displacement field of equation (5.2) yields

$$\{\epsilon\} = [B]\{d\} \tag{5.3}$$

where, $\{\epsilon\}$ are the strain components, and $[B]$ is an element matrix relating the strain components to the node point displacements and consists of derivatives of the interpolation functions. These will also be shown in specific element formulations.

The stress components come from the stress-strain relations that were also given in Chapter 2. Writing them in matrix form results in

$$\{\sigma\} = [E]\{\epsilon\}$$
$$\{\sigma\} = [E][B]\{d\}$$

(5.4)

where, $[E]$ is the material stiffness matrix either for plane stress or plane strain.

Now for any given set of small virtual nodal displacements $\{\delta d\}$ the internal virtual strain energy $\{\delta U_e\}$ is

$$\delta U_e = \int_V \{\delta\epsilon\}^T \{\sigma\} dV$$

(5.5)

where, $\{\delta\epsilon\}$ are the virtual strain components produced by the small virtual nodal displacements, $\{\sigma\}$ are the stress components in the differential material volume at equilibrium, and dV indicates the differential volume element of the continuum.

The external virtual work of nodal forces is

$$\delta W_e = \{\delta d\}^T \{f\}$$

(5.6)

where, $\{f\}$ are the nodal forces. Then using the principle of virtual work

$$\int_V \{\delta\epsilon\}^T \{\sigma\} dV = \{\delta\}^T \{f\} \ .$$

(5.7)

Making the substitutions for strain and stress components from the relations above produces equation (5.8),

$$\int_V \{\delta d\}^T [B]^T [E][B]\{d\} dV = \{\delta d\}^T \{f\}$$

(5.8)

where, $\{\delta d\}$ are small virtual nodal displacements from the equilibrium configuration, and $\{d\}$ are the actual nodal displacements of the material from the unloaded to the equilibrium position.

Since both the virtual and actual nodal displacements are independent of any integration over the element volume we have equation (5.9).

$$\{\delta d\}^T \left(\int_V [B]^T [E][B] dV \right) \{d\} = \{\delta d\}^T \{f\} \qquad (5.9)$$

Canceling $\{\delta d\}^T$ from both sides of the equation yields the element equation

$$[k]\{d\} = \{f\} \qquad (5.10)$$

where the element stiffness matrix is given by

$$[k] = \int_V [B]^T [E][B] dV . \qquad (5.11)$$

This provides the general formulation for any element based on an assumed displacement function that can interpolate for the displacements inside the element from nodal values. The element stiffness matrix then depends on the form of the interpolation functions and their derivatives that create the $[B]$ matrix.

There are two shapes of elements used in two-dimensional analysis: the triangle and the quadrilateral. The two basic element shapes may be linear elements or quadratic elements, where linear and quadratic refer to the order of the assumed polynomial displacement interpolation function used within the element area. The linear triangle is the simplest and was the first two-dimensional element developed. Analysts do not use it much now because it requires many more elements to produce a converged and accurate solution compared with the quadrilateral. However, we still examine its formulation both for academic purposes and for its occasional use in coarse to fine mesh transitions for refining models.

The triangular element illustrated in Figure 5-1 defines an area bounded by the three sides connecting three node points. Within the element area the displacement function is assumed to be of the form in equation (5.12),

$$
\begin{aligned}
u &= a_1 + a_2 x + a_3 y \\
v &= a_4 + a_5 x + a_6 y
\end{aligned}
\qquad (5.12)
$$

where, u and v are displacement components of a material point within the element field, x and y are coordinates of the point, and a_i, $i = 1,2,...6$, are constant coefficients to be determined. This is a linear distribution of the two displacement components for any material point within the element area. The linear function has three undetermined coefficients for each component, and since we have three nodes we may evaluate the three constants by use of the node point values of each component.

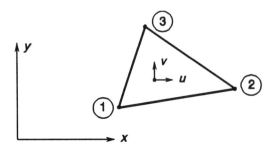

Figure 5-1. Triangular Two-Dimensional Finite Element

Application of the strain-displacement equations given in Chapter 2 to the expressions for u and v illustrate that all three strain components are constant within the element for this assumed displacement field as derived in equation (5.13).

$$\epsilon_x = \frac{\partial u}{\partial x} = a_2$$

$$\epsilon_y = \frac{\partial v}{\partial y} = a_6 \qquad\qquad (5.13)$$

$$\gamma_{xy} = \frac{\partial u}{\partial y} + \frac{\partial v}{\partial x} = a_3 + a_5$$

Also, for homogeneous material throughout the element, the stress-strain relations are all constant, therefore the stress components are also constant.

This displacement formulation also satisfies the compatibility requirements in the theory of elasticity [5.2] for the continuum. The compatibility requirements are that no gaps or overlaps of material may occur during the process of deformation under load. From these equations, we can see that the continuous nature of the function enforces compatibility within the element. On a triangular element, since the interpolation is linear, any edge formed by connecting two nodes that is a straight line before deformation will remain a straight line after deformation. Therefore, any connecting element using the same two nodes for its shared edge satisfies compatibility.

Now the formulation of a triangle element may begin by finding the interpolation function matrix, [N]. If we take the equation for the expression of the u displacement component given in equations (5.12) and evaluate the equation at the three node points, we produce the three equations listed in (5.14).

$$u_1 = a_1 + a_2 x_1 + a_3 y_1$$
$$u_2 = a_1 + a_2 x_2 + a_3 y_2 \qquad (5.14)$$
$$u_3 = a_1 + a_2 x_3 + a_3 y_3$$

Here, u_i are the u displacement components at each node, $i = 1,2,3$, and x_i, y_i are the node coordinates. These equations expressed in matrix form become equation (5.15).

$$\{d_u\} = \begin{Bmatrix} u_1 \\ u_2 \\ u_3 \end{Bmatrix} = \begin{bmatrix} 1 & x_1 & y_1 \\ 1 & x_2 & y_2 \\ 1 & x_3 & y_3 \end{bmatrix} \begin{Bmatrix} a_1 \\ a_2 \\ a_3 \end{Bmatrix} = [A]\{a_u\} \qquad (5.15)$$

Rearranging to express the polynomial coefficients of the u field displacement component in terms of the u nodal displacements gives

$$\{a_u\} = [A]^{-1}\{d_u\} . \qquad (5.16)$$

The inverse of $[A]$ is given by

$$[A]^{-1} = \frac{1}{\det[A]} \begin{bmatrix} x_2 y_3 - x_3 y_2 & x_3 y_1 - x_1 y_3 & x_1 y_2 - x_2 y_1 \\ y_2 - y_3 & y_3 - y_1 & y_1 - y_2 \\ x_3 - x_2 & x_1 - x_3 & x_2 - x_1 \end{bmatrix} \qquad (5.17)$$

with

$$\det[A] = x_2 y_3 + x_3 y_1 + x_1 y_2 - x_2 y_1 - x_1 y_3 - x_3 y_2 \qquad (5.18)$$

which is equal to two times the triangle area.

The equation for displacement component u of equation (5.12) is now equation (5.19) which defines the interpolation functions for the u field displacement component in equation (5.20).

It can be seen that the interpolation functions for the v field displacement component are the same as these above. Then arranging the matrix

$$u = [\,1 \ \ x \ \ y\,] \left\{ \begin{array}{c} a_1 \\ a_2 \\ a_3 \end{array} \right\} = [\,1 \ \ x \ \ y\,][A]^{-1}\{d_u\} \qquad (5.19)$$

$$[N_u] = [\,1 \ \ x \ \ y\,][A]^{-1} = [\,N_1 \ \ N_2 \ \ N_3\,] \qquad (5.20)$$

operations using a conventional order of node degrees-of-freedom (displacement components) yields equation (5.21).

$$\left\{ \begin{array}{c} u \\ v \end{array} \right\} = \left[\begin{array}{cccccc} N_1 & 0 & N_2 & 0 & N_3 & 0 \\ 0 & N_1 & 0 & N_2 & 0 & N_3 \end{array} \right] \left\{ \begin{array}{c} u_1 \\ v_1 \\ u_2 \\ v_2 \\ u_3 \\ v_3 \end{array} \right\} \qquad (5.21)$$

We may now define the element $[B]$ matrix by

$$[B] = [\partial][N] \qquad (5.22)$$

where the partial derivative operator, $[\partial]$, is given by equation (5.23) and results in equation (5.24).

$$[\partial] = \left[\begin{array}{cc} \dfrac{\partial}{\partial x} & 0 \\[2ex] 0 & \dfrac{\partial}{\partial y} \\[2ex] \dfrac{\partial}{\partial y} & \dfrac{\partial}{\partial x} \end{array} \right] \qquad (5.23)$$

By examining the equations which define the interpolation functions and taking the partial derivatives, it can be seen that all the terms of the $[B]$

$$[B] = \begin{bmatrix} \dfrac{\partial N_1}{\partial x} & 0 & \dfrac{\partial N_2}{\partial x} & 0 & \dfrac{\partial N_3}{\partial x} & 0 \\[2ex] 0 & \dfrac{\partial N_1}{\partial y} & 0 & \dfrac{\partial N_2}{\partial y} & 0 & \dfrac{\partial N_3}{\partial y} \\[2ex] \dfrac{\partial N_1}{\partial y} & \dfrac{\partial N_1}{\partial x} & \dfrac{\partial N_2}{\partial y} & \dfrac{\partial N_2}{\partial x} & \dfrac{\partial N_3}{\partial y} & \dfrac{\partial N_3}{\partial x} \end{bmatrix} \qquad (5.24)$$

matrix are constant. The constants, calculated from the values of the node coordinates, therefore remain constant during the integration over the element area. Equation (5.25) now defines the triangle element stiffness matrix, where A is the element area, t is the element thickness, and $[E]$ is the material stiffness matrix for either plane stress or plane strain.

$$[k] = [B]^T [E][B] \int_V dV = [B]^T [E][B]At \qquad (5.25)$$

It is seen that the different material stiffness matrix for plane stress or plane strain leads to a difference in the resulting solution. The differences are primarily in terms involving Poisson's ratio and thus the solutions are usually not much different. In some special cases such as thick wall cylinders with internal or external pressure loading the stress solutions are identical, but the strains will differ. However, in problems involving more than one material or if there are displacement constraints that prevent free motion, then the stress solutions also will differ.

Some of the early finite element programs [5.3] used the triangle element to create a quadrilateral element by subdividing a quadrilateral shape into four triangles using the centroid of the quadrilateral as their apex. After finding the stiffness matrix for each triangle element, assembly of the triangles and condensation of the internal node resulted in the stiffness matrix of the quadrilateral element. This was an effective way to use the triangular element formulation and employ many more elements without tedious input. However, the element of choice now is an isoparametric quadrilateral formulation.

Next we examine the displacement basis for formulation of the isoparametric quadrilateral element. Taig [5.4] developed the element, and Irons [5.5] published its formulation. The quadrilateral element formulation derives from the formulation of a square element. It uses a coordinate system transformation to convert the square to a quadrilateral. Begin with the square element shown in Figure 5-2 with corner nodes.

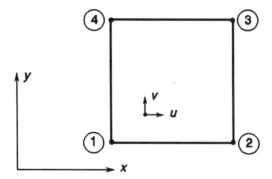

Figure 5-2. Square Two-Dimensional Finite Element

Recognizing that four constants can be evaluated with four nodes, a logical expression for the displacement function components becomes

$$u = a_1 + a_2x + a_3y + a_4xy$$
$$v = a_5 + a_6x + a_7y + a_8xy .$$

(5.26)

Use of the strain-displacement relations here show that

$$\epsilon_x = a_2 + a_4y$$
$$\epsilon_y = a_7 + a_8x$$
$$\gamma_{xy} = a_3 + a_4x + a_6 + a_8y .$$

(5.27)

These equations show that the strain approximation within the element allows ϵ_x to be a linear function of y and ϵ_y to be a linear function of x, while γ_{xy} is a linear function in x and y. So by the addition of one node we have gained a much better approximate solution within the element field for the general case where the strains vary with both x and y throughout the structure domain. Since the stress-strain relations are constant, the stress components may vary similarly within the element field.

Also, in this case the function satisfies compatibility within the element because the function is continuous. Along the element edges for x = constant or y = constant the displacement takes a linear form and thus remains a straight line between any two of the corner nodes. Therefore element connections to other elements satisfy compatibility as long as corner nodes of one element connect to the corner nodes of the adjacent element. Connection of two adjacent elements to a third element such that the edge of the third element spans two of the adjacent elements' edges as shown in Figure 5-3 violates compatibility.

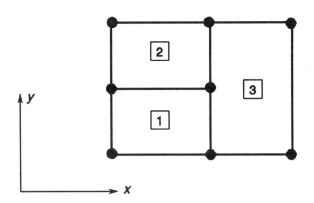

Figure 5-3. Two-Dimensional Element Compatibility Violation

Before continuing the discussion of the square element, make a change of coordinate systems to transform the square into a quadrilateral shape. Square elements can geometrically model very few structures.   A coordinate transformation from x,y to ξ,η produces the quadrilateral element sketched in Figure 5-4.   We call the element an *isoparametric quadrilateral* because the same interpolation functions (parameters) used to define the displacement field define the geometric transformation.

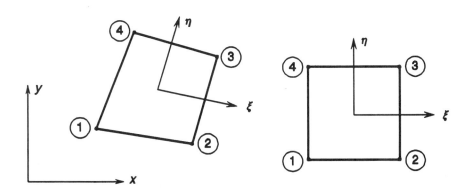

Figure 5-4. Two-Dimensional Quadrilateral Element

The assumed displacement field now is shown in equation (5.28), where the interpolation occurs in the square ξ,η coordinate system. Further, the coordinate ranges of ξ and η are -1 to +1. As before, the evaluation of the equations at the four nodes could produce the set of interpolation

$$u = a_1 + a_2\xi + a_3\eta + a_4\xi\eta$$
$$v = a_5 + a_6\xi + a_7\eta + a_8\xi\eta \tag{5.28}$$

equations by eliminating the constant coefficients in favor of node values. However, this procedure can be made easier by adoption of the Lagrange interpolation formulas [5.6]. Lagrangian interpolation allows the displacement components of any point within the element field to relate to the nodal values by equation (5.29).

$$u = N_1 u_1 + N_2 u_2 + N_3 u_3 + N_4 u_4$$
$$v = N_1 v_1 + N_2 v_2 + N_3 v_3 + N_4 v_4 \tag{5.29}$$

The interpolation formulas, N_i, are given by equation (5.30)

$$N_1 = \frac{1}{4}(1-\xi)(1-\eta)$$

$$N_2 = \frac{1}{4}(1+\xi)(1-\eta)$$

$$N_3 = \frac{1}{4}(1+\xi)(1+\eta) \tag{5.30}$$

$$N_4 = \frac{1}{4}(1-\xi)(1+\eta)$$

Now use the same interpolation formulas to relate the coordinates of any given point x, y in an element to its node coordinates as in equation (5.31). Since we used the same interpolation formula to map the geometry and displacement, we call the element isoparametric (same parameters).

$$x = N_1 x_1 + N_2 x_2 + N_3 x_3 + N_4 x_4$$
$$y = N_1 y_1 + N_2 y_2 + N_3 y_3 + N_4 y_4 \tag{5.31}$$

The element in the x, y coordinate system represents the physical structure, and therefore the calculation of strain requires taking partial derivatives of the displacements with respect to x and y. However, the displacement components are now functions of the square coordinate variables ξ and η. From calculus we know that the derivatives in these two coordinate systems relate through the Jacobian. Using the chain rule for differentiation gives

$$\frac{\partial u}{\partial \xi} = \frac{\partial u}{\partial x}\frac{\partial x}{\partial \xi} + \frac{\partial u}{\partial y}\frac{\partial y}{\partial \xi}$$

$$\frac{\partial u}{\partial \eta} = \frac{\partial u}{\partial x}\frac{\partial x}{\partial \eta} + \frac{\partial u}{\partial y}\frac{\partial y}{\partial \eta}$$

(5.32)

and a similar expression exists for the derivative of the v displacement component. Placing these in matrix form results in equation (5.33) with the Jacobian in equation (5.34).

$$\left\{ \begin{array}{c} \dfrac{\partial u}{\partial \xi} \\[2ex] \dfrac{\partial u}{\partial \eta} \end{array} \right\} = [J] \left\{ \begin{array}{c} \dfrac{\partial u}{\partial x} \\[2ex] \dfrac{\partial u}{\partial y} \end{array} \right\}$$

(5.33)

$$[J] = \begin{bmatrix} \dfrac{\partial x}{\partial \xi} & \dfrac{\partial y}{\partial \xi} \\[2ex] \dfrac{\partial x}{\partial \eta} & \dfrac{\partial y}{\partial \eta} \end{bmatrix}$$

(5.34)

Since we must take the derivatives to define strains with respect to x and y, we can invert equation (5.33) by multiplying by the inverse of the Jacobian as given in equation (5.35).

$$\left\{ \begin{array}{c} \dfrac{\partial u}{\partial x} \\[2ex] \dfrac{\partial u}{\partial y} \end{array} \right\} = [J]^{-1} \left\{ \begin{array}{c} \dfrac{\partial u}{\partial \xi} \\[2ex] \dfrac{\partial u}{\partial \eta} \end{array} \right\}$$

(5.35)

Now we have defined the displacement components, u and v, and the x and y coordinate variables by interpolation formulas which are functions of ξ and η, and the node point coordinates. Combining all these matrix multiplications produces the $[B]$ matrix that relates the strains to node point displacements.

Now to compute the element stiffness matrix we need to take an integral over the element volume. Since the integration involves functions of ξ and η, the variables of integration must be ξ and η. From calculus, the coordinate systems ξ, η and x, y scale through the determinant of the Jacobian such that the differential area is

$$dx\,dy = (\det[J])d\xi\,d\eta \; . \tag{5.36}$$

Then the integral for the element stiffness matrix becomes

$$[k] = \int_{-1}^{+1}\int_{-1}^{+1}[B]^{\mathsf{T}}[E][B]t(\det[J])d\xi\,d\eta \tag{5.37}$$

where, t is the element thickness.

This $[B]$ matrix is a complex function of products and quotients of polynomials in ξ and η, and therefore is not easy to integrate in closed form. The determinant of the Jacobian is also a function of ξ and η, therefore further complicating the integral. To perform this integration, we must resort to numerical integration.

There are many methods for numerical integration of functions, but the method most frequently used in finite element algorithms is Gauss quadrature [5.7]. Our choice for the range of the ξ, η natural coordinate system values from -1 to +1 fits with Gauss quadrature. If we examine Gauss quadrature first in one-dimension integration, the quadrature rule approximates the integral as in

$$I = \int_{-1}^{+1}F(\xi)d\xi$$

$$I \approx \sum_{i=1}^{n} W_i F(\xi_i) \tag{5.38}$$

I is the value of the integral, W_i are weighting factors, $F(\xi_i)$ are values of the integral function evaluated at the n Gauss point locations.

Mathematical handbooks tabulate the Gauss point locations and weight factors for chosen values of n. For example, for one point quadrature ($n=1$) evaluate the function at the coordinate location $\xi=0$, and use a weight factor of 2.0. The approximate integral value is the cross-hatch area in Figure 5-5. For two point integration ($n=2$) evaluate the function at coordinate locations of ±0.5773502692, and use weight factors of 1.0 for the second integral shown in Figure 5-5. Then for three point integration the locations are 0.0 and ±0.7745966692 with corresponding weight factors of 0.8888888889 and 0.5555555556 giving the third integral illustrated in

Figure 5-5.   The Gauss quadrature rules can be shown to give an exact integral value for polynomials of degree *2n-1* when using *n* Gauss points.

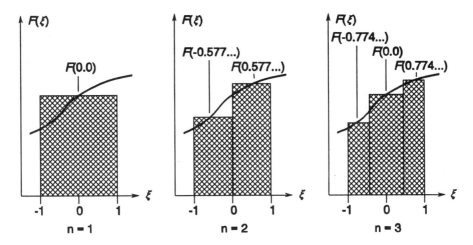

Figure 5-5. Gauss Quadrature

In two-dimensional integration the quadrature rule is

$$I = \int_{-1}^{+1}\int_{-1}^{+1} F(\xi,\eta)\,d\xi\,d\eta$$

$$I \approx \sum_{i=1}^{n}\sum_{j=1}^{m} W_i W_j F(\xi_i,\eta_j) \; .$$

(5.39)

The coordinates of the Gauss point locations are given by the combinations of the one-dimensional locations. For example, one point integration results in the function evaluation at $\xi = \eta = 0.0$. Two by two point integration in two-dimensional space results in the function evaluation at two point locations in each coordinate direction giving a total of four integration points at coordinate values of $\xi, \eta = \pm 0.5773502692$.

This illustrates numerical integration of the element stiffness matrix for the quadrilateral. One point integration is clearly insufficient, but two by two integration is usually adequate. Some programs may choose to use three by three integration for additional accuracy of the integration at the expense of additional numerical computations at the nine gauss point locations.

Finally, numerically integrate the element stiffness matrix by substitution of the values for the ξ, η coordinates for each Gauss point in the product of

$$[B]^T [E][B]t(\det[J])\tag{5.40}$$

and sum the contribution for all the Gauss points.

In either the triangle or quadrilateral element, a higher degree polynomial may approximate the displacement field if we add additional nodes to evaluate the coefficients of the polynomial terms. The triangle expands to six nodes with the addition of a node at the middle of each side of the triangle [5.8]. This allows the displacement function to be approximated by a complete second degree polynomial so that the components are

$$u = a_1 + a_2 x + a_3 y + a_4 xy + a_5 x^2 + a_6 y^2$$
$$v = a_7 + a_8 x + a_9 y + a_{10} xy + a_{11} x^2 + a_{12} y^2\tag{5.41}$$

The strains by application of the strain-displacement relations are then

$$\epsilon_x = a_2 + a_4 y + 2a_5 x$$
$$\epsilon_y = a_9 + a_{10} x + 2a_{12} y\tag{5.42}$$
$$\gamma_{xy} = a_3 + a_4 x + 2a_6 y + a_8 + a_{10} y + 2a_{11} x$$

In the quadratic triangle element formulation the strain components are complete linear functions within the element field, and thus we call the element a *linear strain triangle*. This formulation performs very well and is becoming a popular element for use. It is especially valuable with the emergence of automatic remeshing for reducing solution error, usually called "adaptive meshing" in the literature. The mathematical element formulation may proceed from this point in the same manner as for the constant strain triangle. The resulting element [B] matrix will not be constant over the element area and numerical integration by Gauss quadrature will be necessary.

Adding midside nodes to the quadrilateral allows its displacement approximation function to also be a higher order polynomial [5.9]. Since at least eight nodes (a ninth node at the center for Lagrangian interpolation) are present, it needs an eight (or nine) term polynomial that will include some of the terms from the complete cubic polynomial. As for the corner-noded elements, the quadratic interpolated quadrilateral will provide a little better ability to match the actual displacement variation throughout the element field than the triangle. The element mathematical formulation is conceptually similar to the corner-noded quadrilateral.

Further, if the geometry maps in an isoparametric manner, each side of the quadratic elements may take on a quadratic or parabolic shape. This feature is useful for making a better fit of curved geometrical boundaries of components. However, internal element edges should always remain straight in order to have better numerical accuracy [5.10].

5.2   The Finite Element Model

The model plan should begin by choosing the type of element for use. The linear triangle element can easily develop into a mesh inside almost any arbitrary geometry. However, to produce accurate results there must be many of these elements in the model. In the case studies section, we will see some evidence of the number of triangle elements required versus the number of quadrilaterals to achieve the same accuracy.

In the line element models we have covered thus far, there was little reason for element subdivision other than to define the geometry of the structure. However, two-dimensional cases require element subdivision to achieve an accurate solution. Since element subdivision is required and the exact solution is unknown, a sequence of models with successive mesh refinement is proper. Mesh refinement by further and further subdivision using compatible elements converges to the exact solution. This procedure is known as *h-convergence* because *h* is a common symbol for step size in numerical operations, and its reduction leads to convergence.

Reaching a refined solution by increasing the order of polynomial approximation within the element is another way to achieve convergence. This has become known as *p-convergence*. A direct conversion of a linear element mesh to quadratic elements will yield a more accurate solution. This is the first step in the *p-convergence* method for numerical convergence on the correct solution. The user may easily use *h-convergence* by successive model building in all finite element programs. However, there are usually only linear and quadratic order elements in the element library of most programs that limits the pursuit of *p-convergence*. There are several new commercial codes becoming available now, and a recent text by Szabo and Babuska [5.11] provides good coverage of the *p-convergence* method theory and application.

In some studies, the combination of the two methods leads to the converged solution most quickly [5.12]. Begin the modeling using a linear element, and then refine the model by subdivision until you get a reasonable solution. Follow this by changing to the quadratic element type in the final mesh. If the results do not change dramatically in this last step, then the solution has converged. Some programs make this step relatively simple, so that changing the element from linear to quadratic leads to automatic definition and addition of midside nodes to all elements in the model.

The mesh plan then begins by recognizing the geometry to model and developing of some insight into the expected variation of displacement, strain and stress using approximate engineering calculations. From these estimates, identify the most critical regions in the body and plan for the mesh refinement to correlate with them. You may eliminate all geometric features of relative insignificance with respect to the structural performance. If you choose to use the linear triangle element, remember that the formulation assumes that the strain and stress components are constant within the element and take that into account when developing the element subdivision. Even though the quadrilateral formulation allows some partial linear approximation of the variables, for initial planning purposes, it is better to assume that it also has an almost constant approximation of strain and stress within the element field.

In planning the mesh try to use symmetry whenever possible. The advantages include a reduction of labor of model input, reduction of computer time and cost, and a decrease in computer round-off error in the equation solution since fewer equations exist in the model. There are some drawbacks. Sometimes it becomes more difficult to picture the model. Also, peak stresses may occur along symmetry lines and make it difficult to locate elements properly to show the peak.

Recognize symmetry in two-dimensional objects by observation of geometric patterns that may occur. These may develop by incrementing plane sections, rotating sections about an axis, periodically rotating sections about an axis, or by reflecting a section about a plane. For the symmetric model to provide a solution, the load distribution must also be symmetric on the object. In some cases, we can find solutions for anti-symmetric loading conditions on symmetric objects by proper imposition of displacement boundary conditions.

Displacement boundary conditions enforce symmetry by restricting node points that lie on lines of symmetry to motion along the line of symmetry. For example look at the simply supported beam with central load in Figure 5-6. It has a vertical plane of symmetry at coordinate $x = 0$.

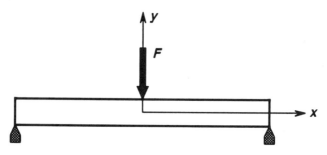

Figure 5-6. Simple Beam with Central Load

When the load applies, the beam will deflect downward and the displacement of every material particle in the right half will be a mirror image of the corresponding particle in the left half. So if the body is symmetric before loading, it is also symmetric after loading. Then we only need to model one half of this beam. If we take the right half, then the outline of the model is in Figure 5-7. The model load reduces by one half because each half of the beam carries its share. The node points that lie on the plane where $x = 0$ are restrained against x direction motion, but left free to move in the y direction.

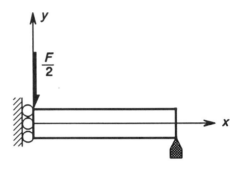

Figure 5-7. One-half Simple Beam Model with Central Load

If the circular disc shown in Figure 5-8 has diametrically opposed loads, it has a horizontal plane of symmetry at $y = 0$ and a vertical plane of symmetry at $x = 0$. Therefore the model should be the quarter section

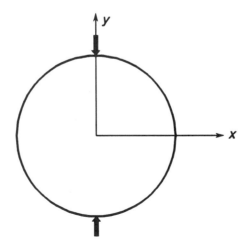

Figure 5-8. Circular Disk with Diametral Load

shown in Figure 5-9. The applied load reduces again by one half. The nodes lying on the vertical plane $x = 0$ are restrained against any x displacement and the nodes lying on the horizontal plane at $y = 0$ are restrained against any vertical displacement.

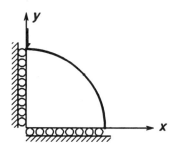

Figure 5-9. One-quarter Circular Disk Model

Displacement boundary conditions provide support and limits to the motion of the model. These should be as accurate and realistic as possible to model the actual behavior of the structure. These limits go at all points where either rigid support or contact with another body restrains motion. Most of all do not apply these restraints arbitrarily or at locations where there is nothing to provide that restraint. After applying all the displacement conditions for support or contact and all symmetry restraints to the model, then be sure that there can be no freedom for rigid body motion of the model.

In a 2-D model, rigid body motion could occur by x translation, y translation, or rotation about the z axis. Examine the displacement restraints applied for symmetry and support conditions, and determine if these restraints will automatically prevent these three rigid body motions. If not, then we need to apply additional boundary conditions to assure that rigid body motion does not occur. If there is a possible rigid body motion and there is a net external force acting on the body in that direction, then we have not recognized all the support restraints since the body is not in static equilibrium. If there is no net external force, then we may select any single node location for restraint to prevent that motion.

For example, in the simply supported beam in Figure 5-6, assume we model the whole beam and the only restraints are vertical fixed displacement at the two support points. Then the body would be free to move in the x direction with a rigid body translation. To prevent this motion select any node and apply one x direction restraint. This one restraint is adequate to prevent rigid body motion. In fact, applying more than one restraint artificially prevents the structure from displacing normally and therefore falsifies the solution. Application of displacement restraints to prevent rigid body motion should not induce any stress conditions in the body.

The applied loads and distributions call for the placement of nodes that will effectively distribute these loadings to the model. The model plan must include an adequate number of node points where the loads apply. Typically, we apply a concentrated force at a single node. This is a common simplification for a surface pressure distributed over a small area. If the actual surface area is smaller than the face area of adjacent elements, then the application of the force at a single node is acceptable. However, if this causes a stress concentration that is too high, then we must distribute the force over two or more nodes at a spacing that limits the boundary stress to the value of expected surface pressure. These considerations also apply when making displacement boundary condition supports at nodes.

Loading may occur by distributed forces such as pressure over large areas, and it must also reduce to node point forces. In two-dimensional models, pressure loads may occur along geometric boundaries of the structure, and therefore must apply to the element edges that lie along the boundary. The resultant nodal forces are a statically equivalent force set that do the same total work on the structure as the pressure load would [5.4]. For linear corner noded elements, the node forces come from simply taking the resultant force acting on an element edge due to pressure and splitting that force between the two nodes on that element edge. Two examples of the node forces that apply for a uniform unit pressure along an edge of unit length are shown in Figure 5-10.

Another load condition that may apply to the body is a distribution of body forces. These are forces generated by action on the material mass

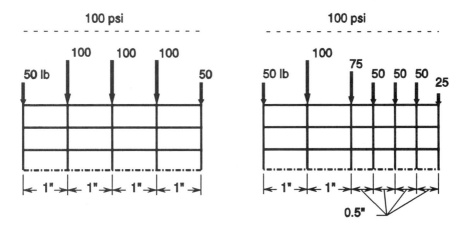

Figure 5-10. Node Forces Resulting from a Uniform Pressure Load

such as in acceleration fields or magnetic fields. In this case for linear elements, the total force on an element volume equally distributes to the corner nodes.

5.3   Computer Input Assistance

At this point, the analyst should have roughly defined the model. In two-dimensional analyses, we want to develop an adequate subdivision within the area defined by the geometrical boundary. For simple geometries and loadings typically a regular array of elements will be suitable and are not very difficult to create with simple replication schemes. However, for more complex geometry involving hundreds or thousands of elements, we need the aid of an area or two-dimensional mesh generator. Most programs provide a mesh generation capability in their preprocessor.

After beginning the preprocessor program, we need to make several selections and check defaults before developing the mesh. These may include the selection of the two-dimensional element type and any options it may have, the type of analysis to be done, and default values for coordinate systems, boundary conditions, material and physical properties. The default coordinate system is usually the cartestian system, but the model might generate better in a cylindrical or spherical system. Then the default may be reset to cylindrical or spherical, and usually, it also may be reset during the progress of model building if desired.

Some programs may have default material property sets or a library of material sets to select from. We should check and be sure that the material properties are suitable for use in the model. If the model uses more than one material property set, be sure to define them before mesh generation begins. The physical property associated with two-dimensional elements is only the material thickness. If more than one material thickness exists in the model, then also define these tables before mesh generation. At this point, the user should have a reasonable idea of what the mesh should be like and may begin mesh generation. In 2-D solid models, it is very helpful to have an automatic mesh generator to map out the area mesh.

The following discussion will focus on several aspects of mesh generation and the types of mesh generators that exist in different programs. The user needs to understand the procedure to control the quality of mesh creation. Of course, any mesh generator requires input of the basic geometry to develop a mesh.

Incrementation methods are the simplest approach for mesh generation. If the model has a long string of nodes with constant node number increment, then by definition of the first and last nodes in the string and the number of nodes in between them, the computer can easily generate the intermediate nodes with either a uniform or biased spacing. Figure 5-11 illustrates this result.

Figure 5-11. Node String Generation

We can replicate the string of nodes using increments of coordinate values and node number to produce additional strings of nodes. This can result in an area node array to define the bounds of a quadrilateral region as shown in Figure 5-12.

Figure 5-12. Node Quadrilateral Array Generation

Within the node array shown in Figure 5-12 we may define either triangle or quadrilateral elements. First define one element, and then generate a row of elements following the first element by incrementing the node numbers of the first element in arithmetic progression. Replicate this row by uniform incrementation of node numbers to define additional rows and complete the element pattern. In this type of mesh generation, the user may control the node numbering to minimize the final bandwidth of the structure. This has been done in the example of Figure 5-13 by using the node number pattern shown.

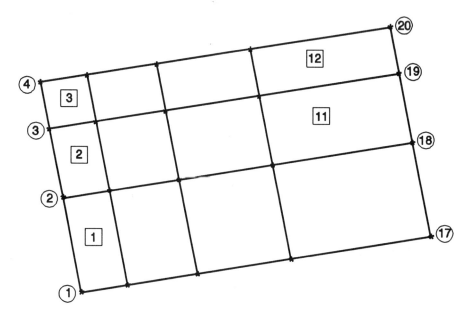

Figure 5-13. Element Generation

If the program uses a wavefront equation solver, then you should arrange the element numbers to minimize the wavefront. Most programs number elements consecutively as it generates them. Then, in the example, the minimum wavefront happens when we generate the short row of elements first and follow by replication of the other rows in the long direction as shown in Figure 5-13. The illustrations emphasize quadrilateral element generation. However, most area mesh generators will produce triangle elements, if requested, by subdivision of quadrilateral elements. It deserves emphasis that the user controls the bandwidth and wavefront of the system of equations in this type of mesh generation by judicious selection of the sequence of node and element generation.

If we can not easily define the geometry by a single quadrilateral space, then many times we can define it by breaking it into quadrilateral segments. We call this a *sides-and-parts* generation method. The geometry in Figure 5-14 may be subdivided into five parts. A program generates the mesh in each of the quadrilateral spaces and then joins the parts by deleting duplicate nodes along the part interfaces. Again the user must exercise control over the pattern of node and element numbering to minimize bandwidth and wavefront.

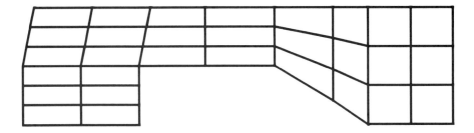

Figure 5-14. Sides-and-Parts Element Generation

Another approach for irregular areas is to perform a coordinate transformation mapping from an approximate fit of squares in an integer geometry to the actual physical geometry. This mapping may be done by laying out a rough equivalent of the actual geometry in an integer space where each square in the integer space corresponds to an element. This procedure is illustrated in Figure 5-15.

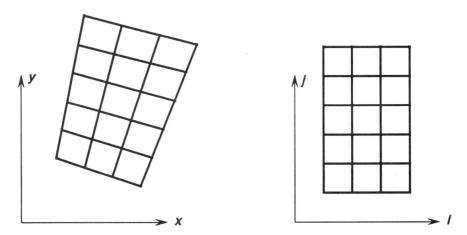

Figure 5-15. Two-Dimensional Mapping Mesh Generation

Sides in the integer geometry correspond to sides in the actual geometry. Sides in the integer geometry must be horizontal or vertical, while sides in the actual geometry can be either straight or curved line segments at any orientation. Correspondence of the closed boundary of the actual geometry with the closed boundary of the integer geometry sets up the mapping. Use of the difference equations, that result from a finite difference approximation of the Laplace equation, determines the interior node locations.

The interior locations result from an iterative solution beginning with a linear interpolation between boundary nodes. Sometimes, in areas of sharply concave boundaries of the actual geometry, linear interpolation of nodes may fall outside the boundary and may not pull back inside the boundary with the iterations. Therefore, the user must examine the generated mesh carefully before proceeding to make sure the mapping was successful. Additional examples of this type of mesh generation are shown in Figure 5-16. In this approach, the user usually has some control over the resulting bandwidth and wavefront by selection of starting location and direction for node and element generation.

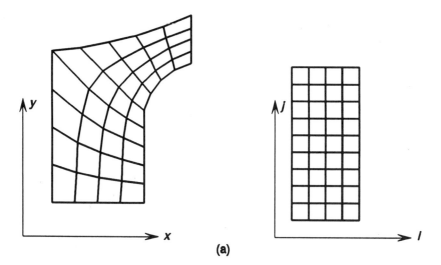

(a)

Figure 5-16. Other Mapping Mesh Generation Examples

An ideal mesh generator, from the users point of view, is one that requires only that the user define the geometry and input the desired element size at key locations throughout the model. This eliminates the planning required to break it into quadrilateral sections or layout an integer map. Some commercial software packages are beginning to feature this type of mesh generator. In this mesh generator, we define the structure boundary using key points, lines, and curves. We input the desired element size for the overall mesh and specify the local sizes at the key points. The computer algorithm then produces a mesh of nodes and elements without further user input as illustrated in Figure 5-17.

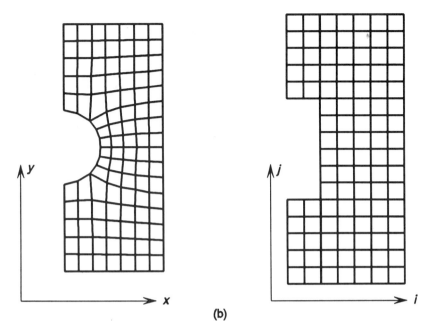

(b)

Figure 5-16. (Continued)

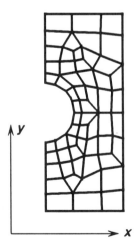

Figure 5-17. Two-Dimensional Free Mesh Generation

In this type of algorithm the user has no real control over the order of generation and therefore cannot guide the program to produce a small bandwidth or wavefront. Additionally the program sometimes produces elements with very poor shape. To reduce bandwidth or wavefront we must use the bandwidth or wavefront renumbering algorithms available within the program. However, if badly distorted elements are present, it is up to the user to reposition some nodes or redefine some elements to fix the distorted elements.

Once a mesh is generated by any method we must check it thoroughly to be sure it is acceptable. The preprocessor should have good graphics display capability to allow the user to plot the mesh and examine all areas of the mesh in detail with pan and zoom capabilities. The boundary geometry that was input should be well fit by the mesh. We can determine this by close examination if the program plots both the geometry definition and mesh simultaneously. In some of the mapping type mesh generators, if a side of the "quadrilateral" corresponds to two or more connected line segments, then sometimes a node will not fall at the line connection and some geometric modeling error will then exist. The user should visually check for high element distortion throughout the mesh. A good mesh also should have a smooth change of element sizes if both large and small elements exist within the model. Nodes that may fall outside the geometric boundaries should also be checked. Look for any improperly defined elements, such as butterfly shapes for example.

After completing and checking the mesh, then begin the application of boundary conditions. In an interactive preprocessor, the normal procedure is to define a displacement condition such as fixing the u displacement component, or fixing v or both, and then selecting the nodes to which this condition applies. For different conditions on different nodes, change the set condition and then again pick those nodes to which the condition applies. A common generation method is to replicate a displacement condition along a row of nodes having a constant node number increment. Some programs allow specification of symmetry planes and then will automatically apply the proper displacement constraints to enforce symmetry.

To apply a boundary restraint that is inclined with respect to the x or y axis, a local node coordinate system may be necessary to orient the constraint. In some programs you may specify the inclined condition by input of the boundary angle with respect to the x axis. During displacement condition input or certainly at the completion of displacement specifications, use the graphic display of the preprocessor to determine visually that you applied the proper restraints and directions.

Load generation is helpful especially when we have a sizable area to load with pressure or body forces (weight or acceleration). Again the normal procedure is specification of a load value followed by the selection of the area to apply it. For point loads, once the load value is set, select all nodes where the load applies. All loads eventually must reduce to node point loads, and most preprocessors can determine the loads from pressure or body force inputs. For input of pressure loading, set the pressure value and then identify the boundary edge where the pressure applies. Nonuniform as well as uniform pressures may be specified.

Body forces such as weight or other accelerations are proportional to the material mass density. Therefore, enter the mass density as a material property, and for acceleration loading, enter the value of linear acceleration or angular velocity. Be careful here to determine the proper units for the quantities entered since some programs have built in constants or default values for mass density and acceleration. The structure is weightless by default unless the user inputs the density and acceleration due to gravity. There should also always be a graphic display of load application to the model for visual proof.

At this stage, the model is complete and ready for analysis. If the bandwidth and wavefront of the generated mesh was not well controlled, depending upon problem size and time requirements, we may need to use a bandwidth or wavefront reduction algorithm. For relatively small problems, say less than 500 degrees-of-freedom, most programs have sufficient solution accuracy. Therefore, if only one case is to be run, then it is not effective to use them. For larger models and models that have several load cases, the user should reduce the bandwidth and wavefront through the renumbering algorithms if they are not already small. Be sure to read the output messages when this process is done to ensure that the bandwidth or wavefront has actually been reduced. In some cases, the algorithm may not be able to determine a more suitable numbering pattern and may actually increase the bandwidth.

5.4   The Analysis Step

In most two-dimensional analyses, there are many more nodes and elements used than in with truss, beam, and frame models. Therefore, there is more potential for error in both the analysis execution errors and overall numerical precision errors. If we checked the model thoroughly in the preprocessor, then we should have caught most execution errors. Execution errors arise by not preventing rigid body motion in the model, improperly defining any element, entering incorrect material and physical properties, and many other factors. The error messages presented by the program usually identify these errors rather easily when they occur.

Numerical precision errors may come about through element distortion, element compatibility violations, and stiffness matrix ill-conditioning caused by large differences in stiffness values of elements [5.7]. Examples of severe element distortion of quadrilateral elements are shown in Figure 5-18. Ideally the elements would remain close to square shape. High aspect ratios, large differences in side length, and very small or large inside angles all contribute to numerical precision errors. In fact, inside angles greater than 180 degrees may cause negative stiffnesses. Most commercial programs will check the element distortion and issue warning messages or cancel execution if the distortion is too high.

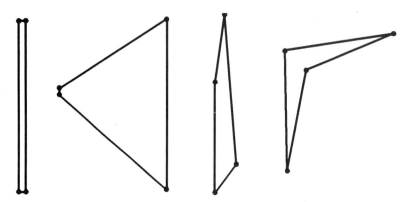

Figure 5-18. Severe Element Distortion

Even though most two-dimensional element formulations guarantee satisfaction of the compatibility requirements in theory of elasticity, modeling errors may still violate compatibility.   Some of these are illustrated in Figure 5-19.

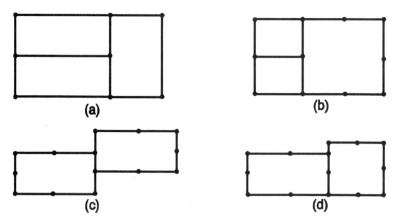

Figure 5-19. Model Compatibility Errors

To satisfy compatibility in corner-noded elements, each element side may only join to a side of one other element. For elements with midside nodes, each must connect to another element at the corners, and all three nodes on a side must connect by use of common nodes.

In most analyses it is good to tailor the mesh to match the solution variations. In areas of slow variation relatively large elements will give good accuracy, while in areas of rapid variation much smaller elements accurately represent the solution better. In the process of mesh refinement, convergence on the solution with this mixture of large and small elements naturally occurs. While this is a reasonable approach to the solution, if the transition from large to small elements occurs too rapidly, it can cause ill-conditioning of the structure stiffness matrix. Ill-conditioning is a result of large differences in stiffness values of adjacent elements connecting to a common node. When the differences become large, the numerical precision of the computer may not be adequate to account properly for both contributions to the stiffness terms of the structure stiffness matrix. These errors may become significant in the sample mesh shown in Figure 5-20.

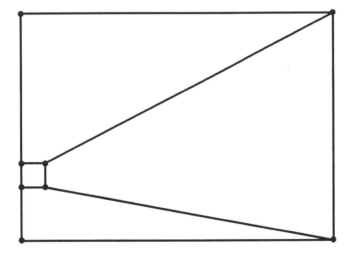

Figure 5-20. Potential Ill-Conditioned Mesh

Also, elements that we usually call structural elements such as beams, plates, and shells may produce ill-conditioning because of large differences between the membrane stiffness and the flexural stiffness in these members. Severe ill-conditioning will be easy to recognize as the results will be obviously wrong. However, less severe cases may produce results that look correct. This places more responsibility on the user to be sure and understand the results of an analysis before accepting and relying on those results.

The numerical performance of the program also will suffer when large models have a large bandwidth or wavefront. This can give large round-off error as the equation solution algorithm progresses. Very few programs to this point provide the user with very much information about ill-conditioning or round-off error. Therefore, the solution to these problems is to develop a good model initially, and avoid the circumstances which cause ill-conditioning and round-off error.

The solution accuracy in two-dimensional analyses is very dependent on the user's ability to evaluate the results and produce a numerically converged solution. In the truss and beam elements, the element formulation was exact and therefore there was no concern about interpolation accuracy. However, these two-dimensional elements require numerical convergence. We accomplish this convergence through careful evaluation of the output and refinement of the model.

5.5   *Output Processing and Evaluation*

Completion of the analysis run will produce a listing file and data files for graphic postprocessing. As mentioned before, scan the printout file for errors in interpretation of input data. The data, especially involving element selection and options, analysis conditions, material and physical properties, and these types of data are relativly easy to scan. Obviously, we can not easily check the lists of node point and element definitions.

Graphic display of two-dimensional data results is necessary to have any chance at making a complete evaluation. Display graphics will be useful for overall checking of model response as well as location of critical areas. This allows examination of the detailed listing file for specific and accurate values in these critical areas. The deformed shape plots have an exaggeration factor adequate to see the deformation. Check boundary condition enforcement, and make a visual judgment as to whether the deformation agrees with the expected response. In some cases, the exaggeration factor may need to be very high to understand fully how the structure is responding. It may help to visualize the shape you would expect if the material of the actual structure was very soft and easily deformed.

Postprocessors normally present stress results in a contour plot form with a range of stress levels. In examining the plots we can make some simple checks. The boundary conditions of the problem require that any stress component perpendicular to a free surface must be zero. Any stress component perpendicular to a pressure loaded surface must equal the pressure value. Where a plane of symmetry exists, the contour lines should approach a normal to the symmetry boundary. There should not be any abrupt changes in contour line direction and it should be continuous. The plot should have expected or understandable shape and location of peak values. These plots result from stress component values at the node points. However, in general, the values at a node common to several elements will not be equal. So there must be some manipulation of the computed stresses to produce these contour plots.

From the element formulation of the linear triangle element, we saw that the assumed displacement field resulted in a constant strain throughout the area. Since the stress-strain relations are constant, the stress components are also constant throughout the element area. If the actual stress field is not uniform, then obviously the stress value must change from one element to the next. Also, in a varying stress field the linear triangle element values are probably most accurate at the centroid of the triangle. This is the value and location that will be given in the printout or listing file for each element. However, the contour plots still need node values. So in use of the triangle element, it is usual to average the element values of all elements common to a node to estimate the node value.

We saw in the development of the linear isoparametric, quadrilateral element that the strain field allowed some component variation within the element area. This variation of the strain components carries over into a similar variation of the stress components. This means that the stress components will have different values at different places within the element. Although we may compute the stress at any point within the stress field including the node points, it has been shown that the values are most accurate at the integration or Gauss point locations for each element [5.13].

Then for the linear quadrilateral element, the stresses computed at the Gauss point locations are the values listed in the printout file. To produce node point values, the Gauss point values extrapolate to the node locations. For any given node, the average of values from all attached elements yield the node value for plotting. However, this involves additional computation and some programs choose to take a simpler approach to get approximate node point values.

Some of the approaches that can be taken for finding nodal values in quadrilateral elements are described in the following paragraphs. The order progresses from least accurate (1) to most accurate (4):

(1) Compute the value at the element centroid. Then assign the centroid value to the four nodes and average the node values from all surrounding elements to find the node values for plotting.

(2) Compute the values at each Gauss point, assign node values from the nearest Gauss point, and then average node values from surrounding elements.

(3) Compute the values at each node in the element, and then average node values from surrounding elements.

(4)   Compute the Gauss point values, pass a plane through the Gauss point values using a least squares fit, extrapolate to the node point locations, and then average node values from surrounding elements.

The approach listed under **(3)** works well for linear quadrilateral elements, but may produce serious errors in the quadratic quadrilateral element with midside nodes. Both the quadratic triangle and quadratic quadrilateral elements should use the procedure listed under **(4)** [5.14]. In most cases a given program will not state which approach it uses, therefore you may want to run your own numerical experiments with simple test cases to determine which approach it uses. This also means that the stress contour plots should be used with caution in determining an accurate value of stress. They are most useful for locating the most critical areas. Then consult the printed output in that area for accurate Gauss point values, and perhaps do a more detailed evaluation of the distribution in that area from these values.

Automatic data scaling normally sets up the data intervals for plotting. The contour algorithm scans the lists of node values to find the data range and selects contour levels to cover the range of values for the given number of contour lines. Most programs also allow the user to select a different number of contour levels and select ranges that can make the plot more effective and clear. The algorithm interpolates along element edges between nodes and locates all points along the edge where contour lines cross. The plotting is then done element-by-element and results in a connected contour throughout the region. If abrupt changes in contour direction are seen on the plot, they will occur at element edges. This is strong evidence that the model does not satisfy equilibrium, and refining the model to yield smoother contour lines should be considered.

Also consider that in contour plotting where there are material interfaces or thickness changes in the two-dimensional model, the stress contours should not necessarily be smooth nor continuous. However, this general approach to contour plotting described above will always produce continuous contour lines. Therefore, the program should allow the user to group all the elements that are of one material or of one thickness and plot the results by groups to avoid artificially smoothing the contour lines.

Upon examination of the stress contours, the user must make some judgment about the validity of the solution. With only results from one model, we can never be sure that we have a converged or accurate solution. The plan is then to produce multiple models using more refined meshes until the solution has converged. The two-dimensional elements presented in this chapter have mathematically guaranteed convergence to the exact solution as the element size approaches zero. It is then appropriate to run multiple solutions with refined meshes to estimate when the convergence has occurred.

Study and observation of the contour plots guides the mesh refinement.

The element size should be smaller in areas where the contour lines are close together (meaning a rapid variation) and larger in areas where the lines are farther apart. There are different measures useful for this evaluation. Typically, researchers developing adaptive mesh refinement have used the strain energy density as a criterion. So if the program you are using will plot strain energy density contours, then they will provide excellent guidance for an improved mesh plan. Another measure of significance in most designs is the Von Mises equivalent stress since it is a good predictor of onset of yield or factor of safety for ductile material structures.

So use of the contour plot of equivalent stress may also provide mesh refinement guidance. Assuming ten contour intervals cover the range from zero to maximum, it would be desirable to have two elements between each interval, in order to have an output resolution of about five percent. At least one element between each interval would provide about ten percent resolution yielding an estimated error of five percent.

Upon reaching the converged solution, the final evaluation of all the analysis results can proceed. The design criteria established at the beginning will determine which quantities to use for final evaluation of the design's performance. For instance, if maximum displacement was of primary concern, then displacement values may come directly from the deformed shape plots or the output listing. If the factor of safety against yielding is of most importance, then compute it by comparison of the material yield strength with the maximum Von Mises equivalent stress found in the analysis.

Although the individual stress components may be of interest, the failure criterion should not be the maximum normal stress theory for ductile materials. Its use can result in serious error when the minimum normal stress is of the opposite sign. The maximum shear stress failure criterion is accurate, but be sure to use the true maximum shear stress and not just the maximum in-plane value. The Von Mises equivalent stress based upon the distortion energy theory is considered to be the most accurate for ductile materials. If the material is brittle or a composite or some other class of material, then the user must determine what failure criterion is proper to use for the given material.

5.6   Case Studies

The first case analyzes a simple cantilevered beam with an end load illustrated in Figure 5-21. The beam has a length to depth ratio of five, which makes it quite short to satisfy the assumptions of elementary beam theory. However, including the shear deformation in the elementary beam theory deflection equation will account for the "shortness" of the beam.

We will compare the finite element analysis with the beam theory deflection at point A and stress at point B.

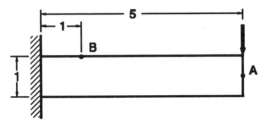

Figure 5-21. Cantilevered Beam with End Load

The beam analysis model first uses triangle elements by beginning with a coarse mesh and refining to converge toward the exact solution. The sequence of triangle element models are shown in Figure 5-22. The first four models are full depth, and the last three models use antisymmetry restraints along the neutral axis with a half-depth model.

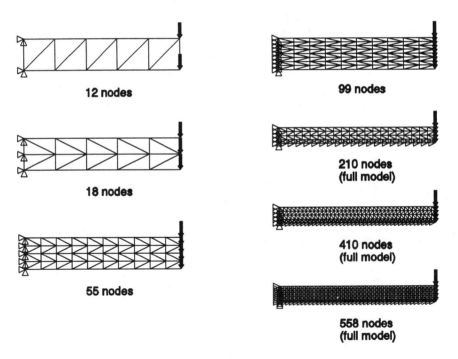

Figure 5-22. Series of Triangle Element Beam Models

Next, the beam model uses quadrilateral elements with the series of models shown in Figure 5-23. The last model of this group is an antisymmetric half-depth model.

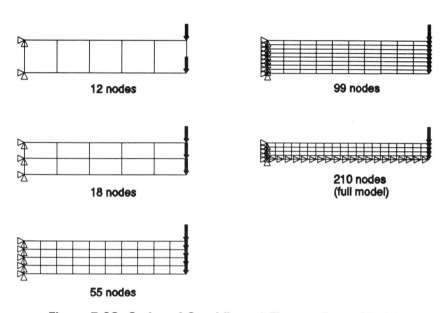

12 nodes

99 nodes

18 nodes

210 nodes
(full model)

55 nodes

Figure 5-23. Series of Quadrilateral Element Beam Models

The results from these analyses are summarized in the next two figures. The displacement of point A at the end of the beam for all the models is given in Figure 5-24. Values for the triangle element models plot with triangle symbols and the quadrilateral element models plot with square symbols. Both types of models produce results that converge monotonically toward the exact solution. However, the quadrilateral element models converge much more rapidly. The 558-node triangle element model is less accurate than the 210-node quadrilateral element model.

The stress convergence comparison is even more dramatic for the quadrilateral element. The graph of bending stress at point B is in Figure 5-25. The stress calculated from the triangle element model is very erroneous even in the 558-node model. The stress calculated in the quadrilateral element model converges very quickly to 99 percent of the theoretical value for the 210-node model.

This case study clearly shows the difficulty in using the linear triangular element in two-dimensional stress analysis problems. Occasionally, a triangular element might be useful in making mesh transitions to accommodate local mesh refinement in quadrilateral element models.

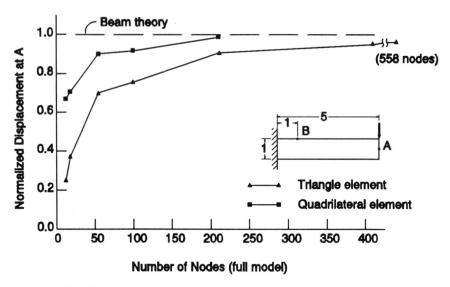

Figure 5-24. Convergence of Displacement at A with Number of Nodes

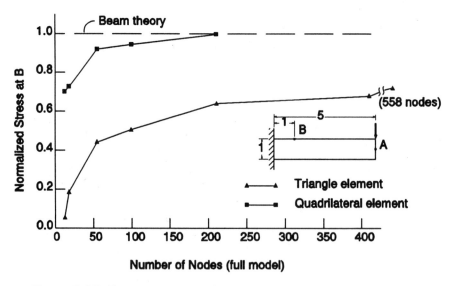

Figure 5-25. Convergence of Stress at B with Number of Nodes

Based on these results, its displacement response may be satisfactory so
it does not seriously degrade the overall model, but stresses computed in
or near the triangular element may have large error. This error should
however be limited to the immediate area.

The second case study analyzes a flat bar in tension with a central hole as a typical stress concentration problem. We will analyze this case with the finite element method and compare the results with the theoretical stress concentration factor. The geometry is shown in Figure 5-26. By taking advantage of symmetry, a one-quarter shaded section of the bar defines the model geometry.

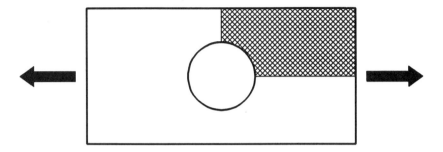

Figure 5-26. Flat Bar with a Central Hole in Tension

The plan for the first finite element model shown in Figure 5-27 has a more refined mesh near the hole because the stress is naturally higher in that area with steeper slopes of change. Displacement restraints apply to the vertical symmetry edge to prevent displacement in the horizontal direction and to the horizontal symmetry edge to prevent displacement in the vertical direction. Node forces calculated and distributed on the right edge provide a uniform stress there of 1 kpsi.

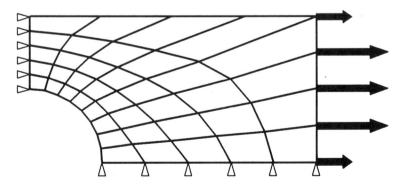

Figure 5-27. First Finite Element Model

After running the analysis, a contour plot of the x-direction stress component appears as shown in Figure 5-28. The contour lines labels correspond to the legend on the left. The element outlines are also shown. The maximum level occurs at the edge of the hole as expected. A clearer view of this region is shown in the enlargement of Figure 5-29. As mentioned before the finite element method gives an approximate, not an exact solution. An estimate of the error in the analysis is the range of stress change relative to the average element value across an element. In this case the corner element includes contour levels from 6 to 9 with a total range of about 1500 psi or about 750 psi from the average. The estimated error is 23 percent.

X STRESS

min	-107.7
0	-102.4
1	333.2
2	768.8
3	1204.4
4	1640.0
5	2075.6
6	2511.2
7	2946.8
8	3382.4
9	3818.0
max	4019.0

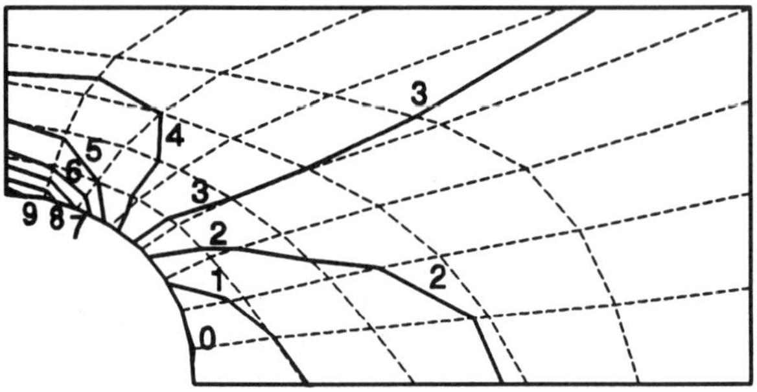

Figure 5-28. Contour Plot of the X̲ Stress Component

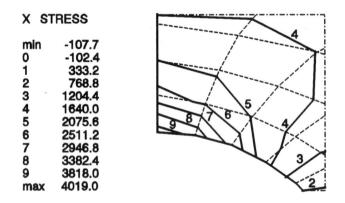

X STRESS

min	-107.7
0	-102.4
1	333.2
2	768.8
3	1204.4
4	1640.0
5	2075.6
6	2511.2
7	2946.8
8	3382.4
9	3818.0
max	4019.0

Figure 5-29. Zoom View of the X̲ Stress Contour Plot

This error margin is large, so we should refine the model. A second model produced was still not sufficiently accurate so we created a third model. This model is shown in Figure 5-30, and it is much more refined around the hole. The zoom view of the x stress component is given in Figure 5-31. We now have a stress range in the corner element of about 700 psi or about 350 psi from the average for a 9 percent estimated error. The maximum value at the edge of the hole is 4300 psi. The nominal, or average stress, on the reduced area section at the hole is 2000 psi which gives a stress concentration factor of 2.15. The theoretical stress concentration factor is 2.18 so the actual error is only -1.4 percent.

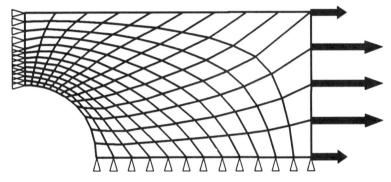

Figure 5-30. Third Finite Element Model

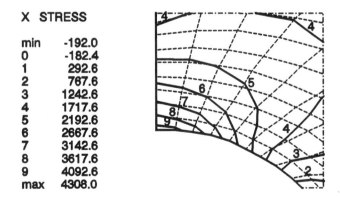

X STRESS	
min	-192.0
0	-182.4
1	292.6
2	767.6
3	1242.6
4	1717.6
5	2192.6
6	2667.6
7	3142.6
8	3617.6
9	4092.6
max	4308.0

Figure 5-31. Zoom View of the $\underline{X}$ Stress Contour Plot (Third Model)

5.7   Closure

The application of the finite element method to two- and three-dimensional problems is where its power is really useful. There are very few closed form solutions to these problems, especially for any but the simplest of geometries. As shown by the case study above, the engineer can reach a very accurate solution by application of proper techniques and modeling procedures. The accuracy is usually only limited by our willingness to model all the significant features of the problem and pursue the analysis until we reach convergence.

Problems

5.1 Show the convergence of finite element models of a simply supported beam with a uniformly distributed load. Refine meshes to determine the maximum displacement and maximum stress within about five percent accuracy. Use a beam length of eight times its height and a unit thickness. Perform the study using each of the following element types if available in the computer program:
(a) linear triangle,
(b) linear quadrilateral,
(c) parabolic triangle, and
(d) parabolic quadrilateral.

   Referring to Figure P5-1 the flexural stress and the center deflection including shear deformation are listed in the following equations.

$$\sigma_x = \frac{3qL^2}{4bc^2}$$

$$\delta = \frac{5qL^4}{16Ebc^3} + \frac{3qL^2(1+\nu)}{5Ebc}$$

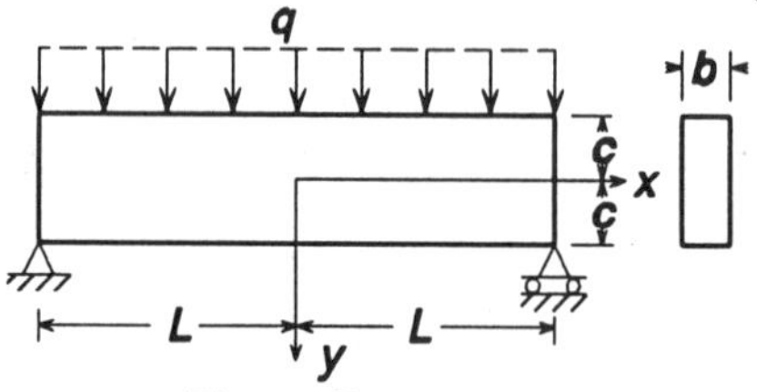

Figure P5 - 1.

5.2 Repeat Problem 5.1 for a cantilevered beam under uniformly distributed load with a 4:1 length to height ratio.

$$\sigma_x = \frac{3qL^2}{4bc^2}$$

$$\delta = \frac{3qL^4}{16Ebc^3} + \frac{3qL^2(1+\nu)}{5Ebc}$$

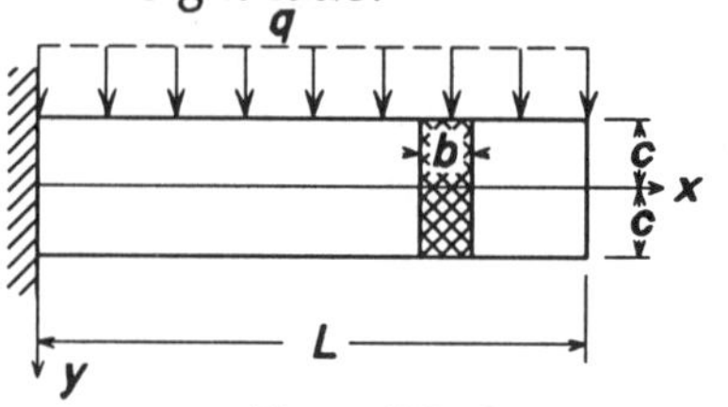

Figure P5 - 2.

5.3 The steel triangle plate in Figure P5-3 bolts to a steel column using ¼-in. diameter bolts in the bolt pattern illustrated. Assuming the column and bolts are very rigid relative to the plate and neglecting friction forces between the column and plate, determine the highest load exerted on any of the bolts. How does this compare with a conventional analysis of a bolt group under eccentric loading? Can you get a reasonable estimate of the stresses in the plate from a simple cantilevered beam analysis? If the plate were made of aluminum, does that change the bolt load? Note: Do not put excessive mesh refinement in the area of the bolt, simply set the displacements of one or a small group of nodes representing the bolt diameter to zero for each bolt. The reaction forces on these fixed nodes are then equal to the bolt loads. However, do use sufficient mesh refinement to get a good resolution of the overall stress distribution in the plate.

5.4 Repeat Problem 5.3 for the bracket bolted with ½-in. bolts to the column in Figure P5-4.

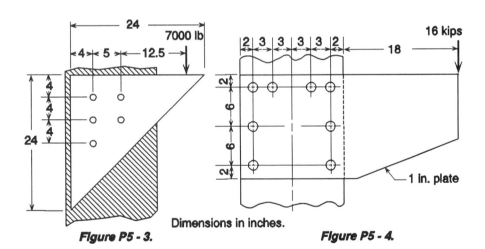

Dimensions in inches.

Figure P5 - 3.   **Figure P5 - 4.**

5.5 A machine member illustrated in Figure P5-5 carries an end load. Simple beam equations furnish the nominal flexure stresses in this member but do not account for stress concentration effects. Determine the stress concentration factors for the two underside radius locations. How do these compare with the values found for similar geometries in many mechanics of materials or machine design books?

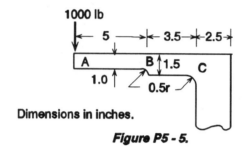

Dimensions in inches.

Figure P5 - 5.

5.6 Analyze the two tensile bar configurations shown in Figure P5-6. Compare your results with published stress concentration factors. Is there a significant interaction effect between the two geometrical discontinuities for the bar in (b)?

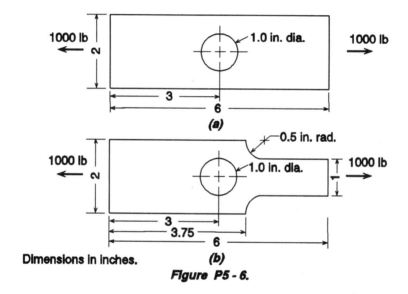

Dimensions In Inches.   *(b)*

Figure P5 - 6.

5.7 Analyze the tensile loaded bar with an off-center hole shown in Figure P5-7. Compare results with the closest published results available. Make successive models with the hole moving closer to the side and see if any pattern develops.

5.8 Determine the stress concentration factor(s) for the notched bar with a center hole in Figure P5-8. Compare with published results for the individual geometries and evaluate any interaction caused by their proximity. You may wish to evaluate additional values of notch radius or varying depths of a constant notch radius.

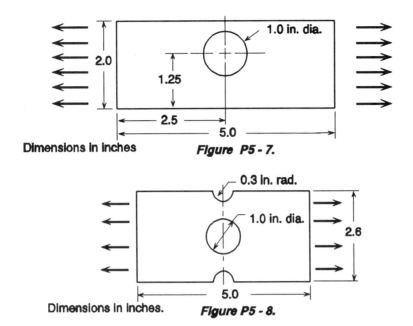

Dimensions in inches   **Figure P5 - 7.**

Dimensions in inches.   **Figure P5 - 8.**

5.9   A ground steel bar and pin with a light drive fit (no play) are drawn
       in Figure P5-9. Analyze both the tensile and compressive loading
       condition through the pin. Locate the maximum stress and stress
       concentration factor for each load condition. If the conditions are
       different, how would that affect the execution of a fatigue analysis
       on the part. Note: Simulate the pin by using an inclined boundary
       condition tangent to the surface for each node along the top half of
       the hole for a tensile load and then along the bottom half for a
       compressive load.

5.10   Determine the stress concentration factors for each end of the
       connecting rod shown in Figure P5-10 when it has tensile loads
       applied through close fitting pins in each hole. Note: Split the rod
       into two models with a tensile load on the cut center section, and
       simulate the pin by using an inclined boundary condition tangent to
       the surface for each node along the load bearing half of the hole.

5.11   A design proposal for the loaded beam illustrated in Figure P5-11(a)
       calls for a hole through the center for passage of some control
       mechanisms as in (b). Does this seriously jeopardize the structural
       performance of the beam? The beam is made of ¾-in. hot-rolled
       UNS G10200 steel plate. Are there other design recommendations
       you would suggest to reduce any structural performance degrada-
       tion?

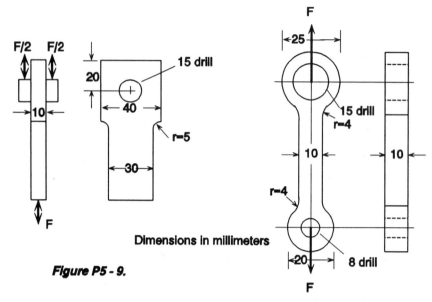

Dimensions in millimeters

Figure P5 - 9.

Figure P5 - 10.

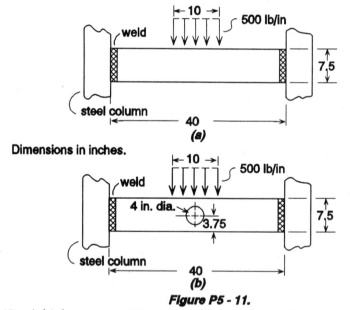

Dimensions in inches.

Figure P5 - 11.

5.12   A high pressure injection molding cylinder has a mounting flange that is drawn in Figure P5-12. It has an internal pressure load on the bore. The design approach for these flanges is to size the cylinder wall thickness based on the equation for hoop stress in thick-walled cylinders. The calculation neglects the bolt holes for the stress on the bore, but then applies a stress concentration factor of 3.0 to the hoop stress calculated on the edge of the bolt hole. The stress concentration factor of 3.0 is for the case of tensile loading of an infinite width plate with a hole. Using a finite element analysis, determine an accurate value of the SCF for the design of the flange shown.

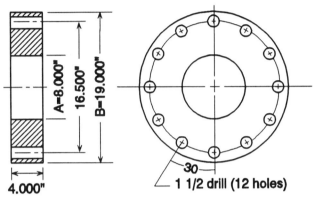

Figure P5 - 12.

5.13 The cross section of an extruded aluminum conduit for hydraulic fluid is given in Figure P5-13. The material is A91100-H14. For an internal pressure of 4 Mpa determine the factor of safety against yielding and the location of highest stress. What would be the effect of increasing the 9 mm dimension?

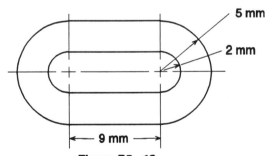

Figure P5 - 13.

5.14 A conventional spur gear with a pitch diameter of 240 mm, a 32 mm bore, and thickness of 12 mm is lightened by boring four holes through the gear thickness. The holes are to be 60 mm diameter on a hole circle diameter of 120 mm as shown in Figure P5-14. The purpose of the lightening holes is to reduce weight and rotational mass moment of inertia. In service the gear must operate up to 5000 rpm. The gear material is UNS C27000 hard-tempered, yellow brass. Is this a safe design change based on the stress produced by centrifugal loading? How much maximum error exists in the pitch radius of a gear tooth caused by radial displacements at the maximum operating speed, and do you think this error has any serious consequence? Are there other design alternatives that you think will work better?

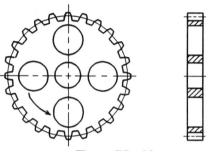

Figure P5 - 14.

5.15 An elongated C shaped section automatic assembly clamp has a preliminary design shown in Figure P5-15. The clamping force is shown in the figure, and the object clamped weighs up to 250 lbs. The clamp and object hang from a pin through the hole at the clamp mid-plane.

Check the design with engineering calculations and a finite element analysis. Decide what material to use and what factor of safety to require for operation of the clamp in a human environment. Recommend any design changes that should be considered or required.

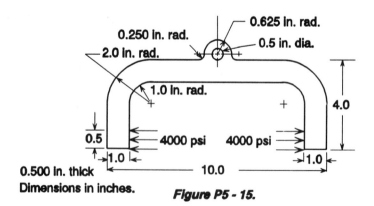

Figure P5 - 15.

References

5.1 Fenner, D. N., *Engineering Stress Analysis: A Finite Element Approach with Fortran 77 Software*, John Wiley and Sons, New York, 1987.

5.2 Timoshenko, S. and Goodier, J. N., *Theory of Elasticity*, McGraw-Hill, New York, 1951.

5.3 Wilson, E. L., and Jones, R. M., *Finite Element Stress Analysis of Axisymmetric Solids*, Aerospace Corporation, San Bernardino, CA., 1967, Air Force Report No. BSD-TR-67-228, (AD-820991, N.T.I.S.).

5.4 Irons, B. and Ahmad, S., *Techniques of Finite Elements*, Ellis Horwood Limited Publishers, West Sussex, England, 1980.

5.5 Irons, B. M., "Engineering Applications of Numerical Integration in Stiffness Methods", AIAA J., Vol. 4, No. 11, 1966, pp. 2035-2037.

5.6 Ball, A. A., "The Interpolation Function of a General Serendipity Rectangular Element", Int. J. Numerical Methods in Engineering, V. 15, No. 5, 1980, pp.773-778.

5.7 Cook, R. D., Malkus, D. S., and Plesha, M. E., *Concepts and Applications of Finite Element Analysis*, Third Edition, John Wiley and Sons, New York, 1989.

5.8 Felippa, C. A., "Refined Finite Element Analysis of Linear and Nonlinear Two-Dimensional Structures," Ph.D. dissertation, Univ. of California, Berkeley, 1966.

5.9 Ergatoudis, I., Irons, B. M., and Zienkiewicz, O.C., "Curved Isoparametric, 'Quadrilateral' Elements for Finite Element Analysis," Int. J. Solids and Structures, Vol. 4, No. 1, 1968, pp. 31-42.

5.10 Taylor, R. L., Simo, J. C., and Zienkiewicz, O. C., "The Patch Test – A Condition for Assessing FEM Convergence," Int. J. Numerical Methods in Engineering, Vol. 22, No. 1, 1986, pp. 39-62.

5.11 Szabo, B. and Babuska, I., *Finite Element Analysis*, John Wiley and Sons, New York, 1991.

5.12 Melosh, R. J. and Utku, S., "Principles for Design of Finite Element Meshes" in *State-of-the-Art Surveys on Finite Element Technology*, ASME, New York, 1983.

5.13 Barlow, J., "Optimal Stress Locations in Finite Element Models," Int. J. Numerical Methods in Engineering, Vol. 8, No. 2, 1976, pp. 243-251.

5.14 Hinton, E. and Campbell, J. S., "Local and Global Smoothing of Discontinuous Finite Element Functions Using a Least Squares Method," Int. J. Numerical Methods in Engineering, Vol. 8, No. 3, 1974, pp. 461-480.

CHAPTER 6

THREE-DIMENSIONAL SOLIDS

Mathematically there is a very straightforward progression from two-dimensional to three-dimensional solids. The element formulation is a simple extension of the approach taken for the two-dimensional case. While the formulation is straightforward, the application of three-dimensional finite element analysis is at least an order of magnitude more difficult. The difficulty lies in the complexity of building a three-dimensional finite element model, which results in a large system of equations that requires significantly more computer resource, and displaying and visualizing the results throughout the volume. Although it is a difficult process, the finite element method provides the only practical approach to solving a fully three-dimensional problem. We can represent all objects in three-dimensional models, but there are some special cases involving symmetry or special geometries that simplify the three-dimensional formulation. These special geometries and their finite element formulations are presented in following chapters.

6.1 Element Formulation

Application of the principle of virtual work developed the general form for the element stiffness matrix in the two-dimensional solid chapter. The form is given in the following equation.

$$[k] = \int_V [B]^T [E][B] dV \qquad (6.1)$$

Thus, we only need to develop the element $[B]$ matrix for any specific kind of element and use the three-dimensional material stiffness matrix, $[E]$, given in Chapter 2 to determine its element stiffness matrix.

The two basic element shapes for 3-D elements are the *tetrahedron* and *hexahedron*. These are similar to the 2-D solid triangle and quadrilateral. The linear tetrahedron is a four node element and the linear hexahedron is an eight node element. These are shown in Figure 6-1. The results and discussion in the 2-D chapter made it clear that the linear triangle element was inferior and rarely used because it requires many elements to produce a converged solution compared with the quadrilateral. This contrast is even more dramatic in the 3-D elements. Even though most commercial programs include the tetrahedron element in their libraries, the finite element users community unanimously condemns it for practical use.

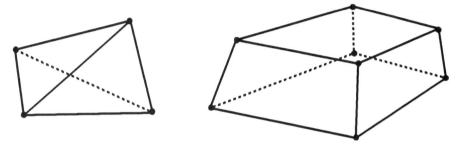

Figure 6-1. Three-Dimensional Tetrahedron and Hexahedron Elements

Three-dimensional models usually require a significantly higher number of elements, and the number required to produce an accurate solution with tetrahedron elements is impractical. Additionally, while the subdivision of a two-dimensional area by triangles is easily visualized, the subdivision of a three-dimensional volume by tetrahedrons is extremely difficult to visualize. In this chapter we present some discussion of the tetrahedron element because it is mathematically simpler to understand. However, only the forms of the displacement interpolation will be given without further development of the tetrahedron element formulation. We develop a more detailed presentation of the isoparametric hexahedron.

The assumed displacement interpolation in polynomial form for the tetrahedron with four corner nodes may be written [6.1]

$$u = a_1 + a_2 x + a_3 y + a_4 z$$
$$v = a_5 + a_6 x + a_7 y + a_8 z \tag{6.2}$$
$$w = a_9 + a_{10} x + a_{11} y + a_{12} z$$

where, u, v and w are displacement components of a material point within the element field, x, y and z are coordinates of the point, and a_i, i = 1, 2, ..., 12 are constant coefficients to be determined. This is a complete linear polynomial distribution of the displacement components throughout the tetrahedron element volume.

Formulation of a quadratic tetrahedron element begins by addition of the six side nodes shown in Figure 6-2. [6.2].

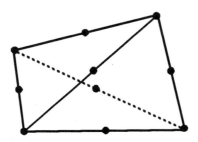

Figure 6-2. A Quadratic Tetrahedron Element

The number of nodes for the quadratic element is now ten, which matches well with the ten terms of a quadratic polynomial in three dimensions. For example, the displacement component, u, is

$$u = a_1 + a_2 x + a_3 y + a_4 z + a_5 x^2$$
$$+ a_6 xy + a_7 y^2 + a_8 yz + a_9 z^2 + a_{10} zx . \tag{6.3}$$

Similar expressions exist for the v and w components.

The quadratic tetrahedron element performs well in constrast to the linear version and it is well suited to automatic three-dimensional mesh generation in complex geometries.   Either of the linear or quadratic displacement interpolations will fit into the proper [B] matrix representation needed for the element stiffness matrix integration.

The linear hexahedron element formulates as an isoparametric element like the two-dimensional quadrilateral [6.2]. In the element's natural coordinate system, ξ, η, ζ, we may assume a displacement polynomial approximation given by equation (6.4), where the eight undetermined constants in each expression evaluate from the eight node point values.

$$u = a_1 + a_2 \xi + a_3 \eta + a_4 \zeta + a_5 \xi\eta + a_6 \eta\zeta + a_7 \xi\zeta + a_8 \xi\eta\zeta$$
$$v = a_9 + a_{10} \xi + a_{11} \eta + a_{12} \zeta + a_{13} \xi\eta + a_{14} \eta\zeta + a_{15} \xi\zeta + a_{16} \xi\eta\zeta \tag{6.4}$$
$$w = a_{17} + a_{18} \xi + a_{19} \eta + a_{20} \zeta + a_{21} \xi\eta + a_{22} \eta\zeta + a_{23} \xi\zeta + a_{24} \xi\eta\zeta$$

As in the two-dimensional case, the equivalent displacement approximation is given by the Lagrangian interpolation functions for the corner-noded element which are listed in equation (6.5).

$$u = \sum_{i=1}^{8} N_i u_i$$
$$v = \sum_{i=1}^{8} N_i v_i \qquad (6.5)$$
$$w = \sum_{i=1}^{8} N_i w_i$$

The interpolation formulas, N_i, are given by equation (6.6). The node arrangement and natural coordinate system for this element are shown in Figure 6-3.

$$N_1 = \frac{1}{8}(1-\xi)(1-\eta)(1-\zeta)$$

$$N_2 = \frac{1}{8}(1+\xi)(1-\eta)(1-\zeta)$$

$$N_3 = \frac{1}{8}(1+\xi)(1+\eta)(1-\zeta)$$

$$N_4 = \frac{1}{8}(1-\xi)(1+\eta)(1-\zeta)$$

$$(6.6)$$

$$N_5 = \frac{1}{8}(1-\xi)(1-\eta)(1+\zeta)$$

$$N_6 = \frac{1}{8}(1+\xi)(1-\eta)(1+\zeta)$$

$$N_7 = \frac{1}{8}(1+\xi)(1+\eta)(1+\zeta)$$

$$N_8 = \frac{1}{8}(1-\xi)(1+\eta)(1+\zeta)$$

To map the cube in the natural coordinates, ξ, η, ζ, to a hexahedron in x, y, z coordinates based on the isoparametric formulation, use the same interpolation formulas to define the coordinates of any given point x, y, z in an element from its node coordinates as in equation (6.7).

The further development of the linear hexahedron element follows the same steps used in the two-dimensional isoparametric quadrilateral

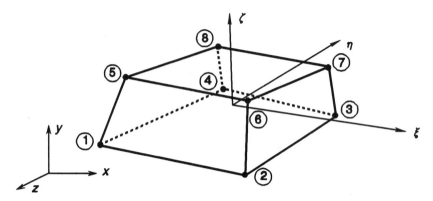

Figure 6-3. A Linear Hexahedron Element

$$x = \sum_{i=1}^{8} N_i x_i$$

$$y = \sum_{i=1}^{8} N_i y_i \qquad (6.7)$$

$$z = \sum_{i=1}^{8} N_i z_i$$

element. Define the Jacobian, use the derivative operation defined in Chapter 2 relating strains to displacement to create the element $[B]$ matrix, and then numerically integrate using Gauss quadrature to get the element stiffness matrix.

$$[k] = \int_{-1}^{+1} \int_{-1}^{+1} \int_{-1}^{+1} [B]^T [E][B] (\det[J]) \, d\xi \, d\eta \, d\zeta \qquad (6.8)$$

Of course, 2 point quadrature requires 2 x 2 x 2 or 8 evaluations of the integrand and 3 point quadrature requires 27 evaluations to find the stiffness matrix. Obviously, the computational effort is increasing rapidly in three-dimensional analysis [6.3].

Further, to form a quadratic hexahedron element requires at least 12 additional side nodes, making 20 nodes in all, for the 3-D solid hexahedron element. The 20 node element is shown in Figure 6-4. Complete Lagrangian quadratic interpolation requires 27 nodes with nodes positioned on the midface of each side and a node at the element center. Therefore, the user may find either a 20 or 27 node element in different finite element code libraries, but the 20 node element is most common in commercial codes.

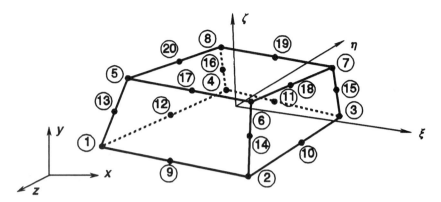

Figure 6-4. A Quadratic Hexahedron Element

6.2   The Finite Element Model

In most 3-D models the linear hexahedron is the most frequently chosen element for building the model. As stated before, the linear tetrahedron is unwieldy because of the huge number of elements required to produce a good solution.   So, if the geometry of the structure or the mesh generation program works better with a tetrahedron element, the user should opt for the quadratic tetrahedron. There may be some areas of a structure where using a tetrahedron may more easily model the geometry, or some locations where element size transitions must occur. In either event, any area modeled by the linear tetrahedron element should have its results very carefully evaluated for convergence.   Use of the linear hexahedron element for initial models will allow conversion to the quadratic hexahedron for the final check of convergence of results.

Because of the increased work and complexity of three-dimensional models, the analyst should always devise some appropriate two-dimensional approximations to the problem. Perform two-dimensional analyses on these cases first. This helps determine where the locations of concern are in the three-dimensional model as well as providing valuable insight into the variations which occur and the number of element subdivisions that might be suitable in the 3-D model. These 2-D cases can be run much quicker and thus may eliminate much of the consecutive or adaptive modeling required to produce a good 3-D solution.

In developing a mesh plan, begin by recognizing expected variations in displacement, strain, and stress and plan the element subdivision accordingly.   In 3-D models, element subdivision is required in every dimension of the structure so the mesh can very quickly become difficult to picture. This is where the choice of hexahedron elements helps because it is easier to see a three-dimensional structure made of "bricks" than one made of "pyramids".

In three-dimensional analysis it is even more important to recognize and use symmetry conditions as much as possible. The 3-D model very rapidly increases the number of system equations and computations required to achieve a good solution. The use of symmetry will help keep the analysis practical.

Symmetry in three-dimensional objects must always occur by reflections about a plane except in the special case of axisymmetric bodies. As in two-dimensional cases the geometric symmetry must couple with load symmetry or load anti-symmetry for symmetric models to work. In 3-D models imposing displacement boundary conditions so node points lying on any plane of symmetry can only move within that plane enforces symmetry. For example, look at the three-dimensional cantilevered beam with loads on the outer corners shown in Figure 6-5.

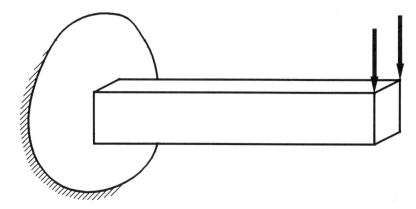

Figure 6-5. A 3-D Cantilevered Beam with Corner Loads

There is a vertical plane of symmetry passing midway through the structure in the x-y plane so the 3-D model only has to include half the beam as illustrated in Figure 6-6. All the nodes that lie on the x-y plane at $z = 0$ have their z displacement component fixed.

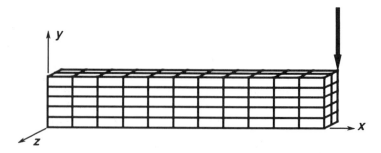

Figure 6-6. A Symmetric Model of the Beam with Corner Loads

A sphere with diametral loads shown in Figure 6-7 has a plane of symmetry in the *x-y* plane, the *x-z* plane and the *y-z* plane so the 3-D model can use an octant of the sphere. However, in this case, there is further symmetry in that any plane passing through the *y* axis is also a plane of symmetry. This allows the model to form a symmetric wedge as illustrated in Figure 6-8. Also, this problem can use axisymmetric elements presented in the following chapter.

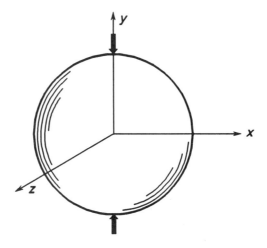

Figure 6-7. A 3-D Sphere with Diametral Loads

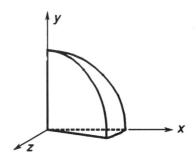

Figure 6-8. A Symmetric Wedge Model of the 3-D Sphere

In all 3-D models displacement restraints apply where supports exist or to enforce symmetry conditions. Following specification of all restraints for these conditions, we also must prevent any remaining possible rigid body motion of the model. In a 3-D model rigid body motion can occur by x, y, or z translation or rotation about the x, y, or z axis so there are six rigid body modes. In any model we must examine the displacement restraints already applied and determine if this set of restraints will prevent all six rigid body modes of displacement. If not, then apply only enough additional restraints to prevent the motion. These may be arbitrarily placed if the structure is in equilibrium. However, do not over-restrain the structure or place restraints to restrict its ability to respond freely to the applied loads.

Plan the three-dimensional model to properly position and apply the structure loads. Application of a concentrated force to a single node causes the same concerns in 3-D models as it did in 2-D models. The elements sharing that node should be large enough that the area of influence of the concentrated force does not significantly affect the solution in other areas of interest.

Pressure loading on any face of linear elements results in the obvious distribution of one-fourth the total force produced on the face area by the pressure to each node. However, in the quadratic element with midside nodes, the distribution of node forces from a uniform applied face pressure results in an apparent illogical pattern of distribution [6.2]. The distribution to produce a uniform normal stress on that face of the element relies on the equivalent work consistent with the interpolation functions on the given face. The resulting distribution for a uniform tension applied to a face with area, A, is shown in Figure 6-9.

Similarly, the body forces such as weight or inertial loads equally distribute to all the nodes in a linear element, but we must interpolate them properly for the parabolic element. Computation of the resulting nodal forces is normally done in the program preprocessor, but the user should know that a set of equal forces on the nodes of a meshed area does not yield uniform pressure or stress.

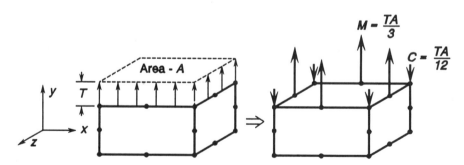

Figure 6-9. Node Forces From Uniform Pressure on a Parabolic Element

6.3   *Computer Input Assistance*

Before beginning any computer program's preprocessor the analyst should prepare a well-planned initial model. Except for the simplest of 3-D cases, the use of a 3-D mesh generator is required. Mesh generation in 3-D structures is obviously much more involved than in 2-D. Here the user must define the 3-D object by combinations of points to define lines, lines to define areas, and areas to define volumes. Some preprocessors may operate interactively with or accept input from a geometric solid modeling program. In this case the geometry description comes from input to the solid modeler.

Most of the mesh generation programs operate on user defined volumes. The mesh within a volume usually follows a mapped approach similar to the 2-D case. This requires the volume to be a hexahedron shape although the edges may curve [6.1].

Complex geometries may require more than one hexahedron volume to describe its geometry completely. All the individual volumes that compose the total geometry must have mating surfaces so the resulting 3-D volume mesh is compatible throughout. A couple of examples of 3-D meshes generated using this approach are illustrated in Figure 6-10.

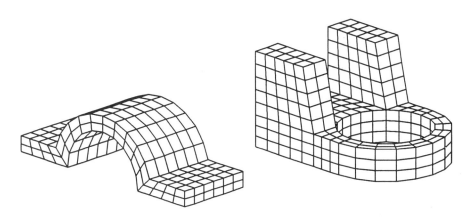

Figure 6-10. Example Meshes of 3-D Solid Elements

In 3-D models defining the surface geometry and simply specifying desired element sizes at key locations is even more appealing to the user than in 2-D models. In this way the user does not have to do all the planning and visualization exercises required to create hexahedron volumes and make sure all the partial volumes fit together. An example of this type of mesh generation appears in Figure 6-11. While this makes

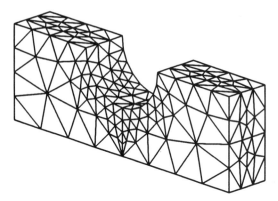

Figure 6-11. Free Mesh Generation of 3-D Solid Elements

the burden on the analyst much lighter, visual examination of some of the element faces show that much element distortion can occur in this type of generator. Further, if you observe the graphics display of the elements in progress you can see the irregular sequence in which elements and nodes are drawn. The irregular sequence produces a very large bandwidth and wavefront. This type of mesh then must use an optimizing algorithm to reduce bandwidth or wavefront to reduce roundoff error and solution time.

Mesh checking becomes a very important step since so much of the model and so many of the element edges are hidden from view in any graphics display that is comprehensible. One of the features of many pre-processors is a free edge check that will show if any internal elements have free edges, i.e. a gap within the material, because the adjacent elements do not connect. Element distortion checks determine if distortion of any element is too severe to spoil program execution. Complete the mesh checking and renumbering by the bandwidth or wavefront optimizer before proceeding with the boundary condition and load specification.

The displacement restraints enforce symmetry and represent the structure supports. In 3-D models the capability of a preprocessor to accept a specified symmetry plane and then to determine which nodes lie on this plane is invaluable. Usually, we set the restrained displacement components before selection of nodes for restraint. In a general purpose 3-D finite element program with all types of elements, the program will normally have three translation and three rotation components available at a given node. Thus, there are potentially six degrees-of-freedom available at every node. However, the 3-D solid element formulation only uses the translational components, so it is only necessary to specify the translation restraints. Most programs will automatically drop all degrees-of-freedom not used in the chosen element's formulation.

Application of loads involves input of a value set followed by selection of locations where the loads apply. Apply individual node force components by a straight forward listing of node numbers or by node picks from the graphics screen. Apply pressures loads to element faces. If there is a sizable area of pressure application, it can become rather tedious to pick each element. Some programs offer the ability to specify a surface or plane for pressure application, and the program then scans all the elements to determine which ones have a free face on the given surface or plane. Again examine carefully the graphic display of the result of that action to be sure the computer algorithm did not make a mistake.

6.4 The Analysis Step

Most 3-D models will naturally have a large number of system equations to solve. Obviously, this increases the chances of model errors through improper element definitions, improper boundary conditions, and other input errors primarily because we can not easily display the model graphically in full detail. Also, the sensitivity of the numerical performance to the large bandwidth or wavefront increases, and any abrupt element size changes will more quickly cause ill-conditioning of the structure stiffness matrix.

The chance of ill-conditioning is further increased because of difficulty in developing a fully 3-D mesh without incorporating large element aspect ratios. Many 3-D models will have sections that are somewhat thinner than other sections of the model. Thus, to get a reasonable subdivision in the thin areas, the user must employ a large number of elements in the thicker dimensions or else use elements with high aspect ratios.

The naturally large bandwidth required for 3-D models places a heavy burden on the numerical algorithm used for equation solution. The large bandwidth significantly increases the chance of round-off error in these algorithms. Neither the round-off errors nor errors due to ill-conditioning are easily detectable (unless the output is total nonsense). This increases the importance of doing approximate 2-D analyses of the problem to provide the analyst with some insight into the expected results.

6.5 Output Processing and Evaluation

Upon completion of the analysis run, the resulting listing files and data files will normally be very large, making any practical scanning of these data out of the question. Therefore, the analyst must turn to the graphic presentations. Engineering judgment and intuition can aid the engineer a great deal in evaluating 2-D responses because they are relatively more easily visualized. However, in 3-D applying our judgment and intuition becomes more difficult.

A careful study of the graphic display of an exaggerated deformed shape of the model can help us interpret how the model is responding to the loads. For the graphic display to be at all sensible, it is ordinarily done with hidden line algorithms such that only the element mesh visible from the selected point of view shows. Because of this, we must take several views from different directions to develop an understanding of the deformation. The goal in the study of deformed shape is to satisfy ourself that we have modeled the object correctly. Do this by studying the deformed shape and reconciling that shape with our knowledge or judgment of what the deformed shape should be. The interaction effects of stresses and strains in 3-D generate many unusual behaviors which are difficult to imagine in advance. We may observe such an occurrence when we look at a 3-D solid beam in the case studies.

The stress results are usually in the form of contour plots on the visible surface. The analyst should again evaluate the displayed stress values compared with the stress boundary conditions at free surfaces or pressure loaded surfaces. When studying the stress contour displays most computer programs will list the minimum and maximum values found within the whole model, so if these values are not seen in the current view then other views should be taken. In some cases the peak stresses may occur at internal points that are not visible on the display. However, most of the time the peak stresses occur on some surface, so we usually find them by observation of surface plots. Most programs also provide a capability to slice through the object and display the resulting cut surface with its stress contours.

As in 2-D structures, the stress plots also result from computation of node point values from the values computed within each element and then averaged at the nodes. Since the node values in this case must extrapolate in three directions, the potential for extrapolation error is larger than for the 2-D cases. This means that the analyst must be even more careful in accepting values from the displayed contours. After determining critical locations, the analyst may then return to the printed output or listing file and check the stress results. The program usually prints results within the elements at the Gauss point locations, and the results are most accurate numerically at those locations. From study of the individual element results, the engineer's judgment about the extrapolation is probably better than the algorithms employed to create the displays.

Following evaluation of the first model, appropriate mesh refinement should be done to improve the model and generate the next set of results. Because of the difficulty of developing and running 3-D models this process should be done very carefully to limit the number of iterations required. Development of appropriate 2-D models prior to the 3-D analysis usually helps. Since the potential for round-off error and ill-conditioning increases rapidly as mesh refinement is done the analyst must proceed cautiously.

6.6   Case Studies

A simple cantilevered beam is chosen to demonstrate three-dimensional finite element analysis with 3-D solid elements. Although it is a simple problem, some interesting aspects of engineering mechanics of solids emerge that are not seen in the normal one-dimensional or two-dimensional analysis of beams. The beam is short and wide which violates the assumptions made in simple beam theory, but we will evaluate the ability of simple beam theory to make predictions of the beam's displacement and stress response to an end load. The beam is 5 in. long by 4 in. wide by 1 in. thick. The clamped end is on the *y-z* plane with the *x* axis running along the length. A half-model of 3-D linear solid hexahedron elements is drawn in the hidden line plot of Figure 6-12. The model is fixed on the *y-z* plane, has symmetry restraints of $v = 0$ on nodes lying on the *x-z* plane, and has a line load totaling 1000 lbs. on the top nodes of the end plane.

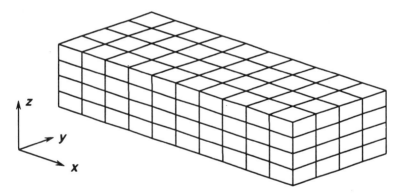

Figure 6-12. Finite Element Model of a 3-D Cantilevered Beam

After running the solution, first examine the deformed shape plots to check for proper boundary condition placement and reasonableness of the deformed shape. A hidden line view along the *y* axis is given in Figure 6-13. The outline of the undeformed geometry is shown with a dashed line. The end displacement and gentle curved shape up to the fixed condition at the wall appear reasonable. The concentration of close lines showing at the bottom means that additional element faces other than

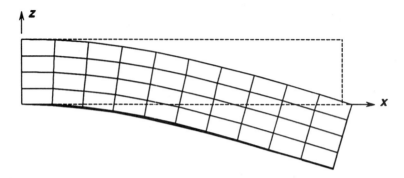

Figure 6-13. Side View of Beam Deformed Shape

the ones in direct view are also showing. The question to ask is "Is this a reasonable deformation?."

In order to check this out another view is taken along the x axis at a location 1 in. out from the wall where this distortion looks to be highest. This view is seen in Figure 6-14 using a much higher exaggeration factor. At this cross section the symmetry restraints on the x-z plane are correct. The outside edge face is rotating as the beam edge deflects upward relative to the overall downward beam deflection. This seems like very strange behavior. Now it is up to the analyst to either explain the behavior or determine what is wrong with the model.

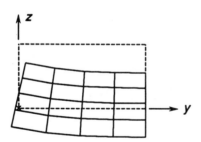

Figure 6-14. Axial View of the Deformed Shape Cross Section at $x = 1$

The behavior is explainable. The material strain created through the effect measured by Poisson's ratio causes the behavior. In the cantilevered beam with a downward end load the axial stress is compressive below the neutral axis and tensile above the neutral axis. The axial compressive strain under the neutral axis causes the material to expand laterally by an amount equal to Poisson's ratio times the axial strain. The axial strain is linearly proportional to the distance from the neutral axis (for simple beams). Therefore, the expansion varies linearly from a maximum at the bottom surface to zero at the neutral axis and contracts above the neutral axis due to tensile axial strain.

Now examine the stress results. Based on simple beam theory the x component of normal stress should be of most concern. Taking a side view of the beam, the stress contour plot appears in Figure 6-15. Based on the contour values and intervals we expect that the space from the highest contour to the surface and the lowest contour to the surface would be about the same as the other spaces between contours, yet the spacing near the surface looks too small.

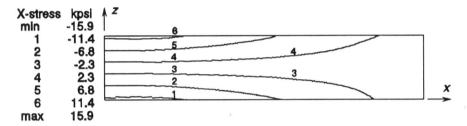

Figure 6-15. Axial Stress Contours on the Side of the Beam

Upon further examination, these stresses increase toward the center of the beam as shown in the cross section in Figure 6-16. This shows that the axial stresses are higher on the mid-plane and decrease toward the side.

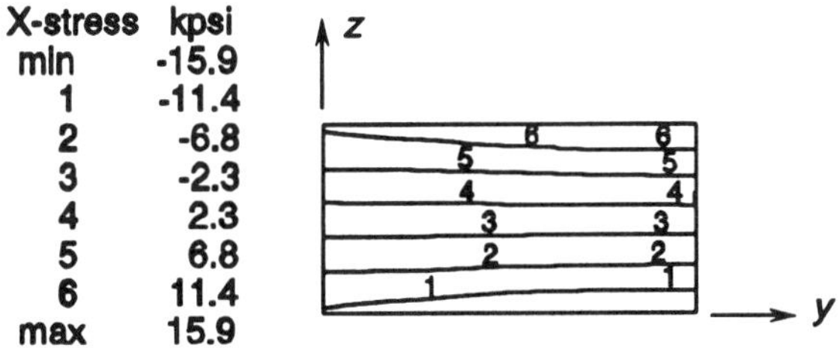

Figure 6-16. Axial Stress Contours on the Section at x = 0

Checking the other stress components reveals a significant state of stress for the y component at the wall support. These contours are in Figure 6-17. While they are less than the axial stress values, they are significant at or about 33 percent of the axial value and caused by the Poisson ratio strain.

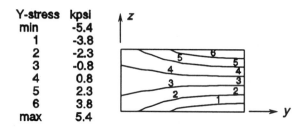

Figure 6-17. Lateral Stress Contours on the Section at x = 0

These components combine to form the Von Mises equivalent stress plotted in Figure 6-18 on the mid-plane of the beam. The maximum value occurs slightly out from the wall.

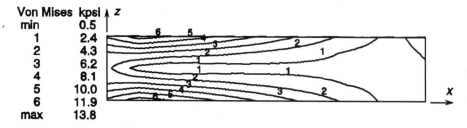

Figure 6-18. Von Mises Equivalent Stress Contours at x = 0

Now compare the results calculated from beam theory with these results. From beam theory the end deflection is 0.00833 in. and the finite element result is 0.00835 in. The maximum flexural stress from beam theory is 15.0 kpsi and the finite element result is 15.9 kpsi. These are amazingly good correlations, showing that elementary beam theory works pretty well even for short beams. However, the case study shows that some additional important things are not given by beam theory.

Problems

6.1   Determine the maximum stress component values, the maximum Von Mises equivalent stress, and the deflections of the loaded corners for the steel cantilever beam in Figure P6-1. Determine the values first using conventional beam and torsion equations and then by doing a finite element analysis. How do the results compare?

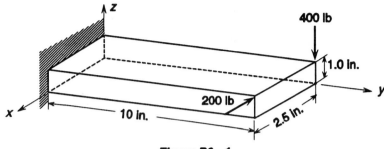

Figure P6 - 1.

6.2 Choose a suitable steel material for the foot pedal lever design in Figure P6-2. Base the initial selection on results from conventional beam equations and then verify or change the selection based on results from a finite element analysis. Suggest any design modifications you think should be made. Do the deflections seem reasonable? What design modifications would reduce the deflections?

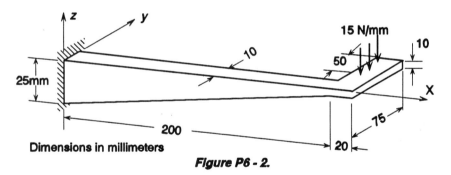

Figure P6 - 2.

6.3 The steel machine member in Figure P6-3 was analyzed in a two-dimensional model in Chapter 5 Problem 5.5. Do a three-dimensional analysis and compare the resulting stress concentration factors to the two-dimensional values.

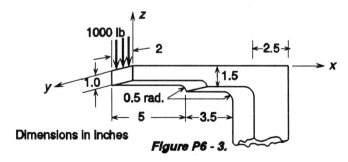

Figure P6 - 3.

6.4   The cast iron machine tool jig in Figure P6-4 has side slots for
locking pins. For the given end load find the maximum stress and
location and the maximum displacement.   How do the values
compare with conventional beam equations results?

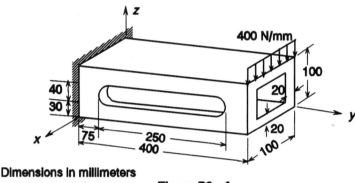

Dimensions in millimeters

Figure P6 - 4.

6.5   A rotating shaft in Figure P6-5 is shown with lateral loads that
produce flexure stress on the central section. There is obvious stress
concentration at the grinding relief groove. The relief groove is only
2.5 mm deep.   Determine the stress concentration factor for the
given geometry under flexure loading and compare with any
published values you can find for the same or similar geometry.
You may use the same model to find the stress concentration factors
for axial and torsion loading.

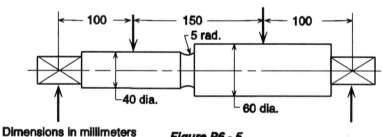

Dimensions in millimeters   *Figure P6 - 5.*

6.6   An alternative way to that shown in Problem 6.5 to provide a
grinding relief groove in a shaft is represented in Figure P6-6.
Determine the stress concentration factor due to the flexure loading
shown and compare with any published values you can find for the
same or similar geometry. You may use the same model to find the
stress concentration factors for axial and torsion loadings.

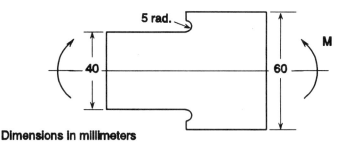

Dimensions in millimeters

Figure P6 - 6.

6.7 The triangular loading ram plate in a bearing race quench press drawn in Figure P6-7 must satisfy three design criteria when subjected to the eccentric reaction forces. The three criteria are σ_{allow} = 30 kpsi, maximum out-of-plane displacement = 0.004 in, and maximum plate rotation = 0.02 degrees. The plate slides vertically along the columns through ball bushings with maximum applied forces of F_A = 3600 lb. The reaction is located at point R. Does the design shown meet the criteria? If we assume it does not, what design changes would help?

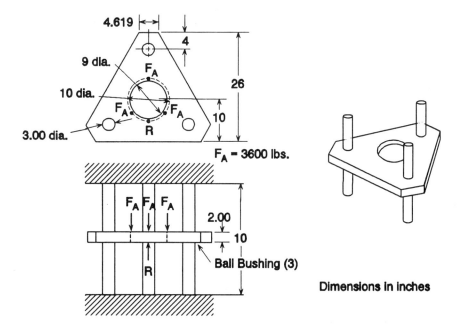

Figure P6 - 7.

6.8   Using standard spur gear tooth geometry, select the parameters for a gearset and do a 3-D finite element analysis of meshing teeth as illustrated in Figure P6-8. For simplicity in model building you may use the same mesh for both teeth. Examine the contact stress and tooth root stress concentration factor. Compare with conventional design relations used in machine design texts or from the AGMA (American Gear Manufacturers Association).

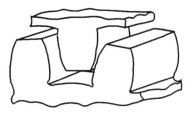

Figure P6 - 8.

6.9   Evaluate the flywheel design in Figure P6-9. The flywheel made of an aluminum alloy A03190 casting operates up to 5000 rpm. Your evaluation should conclude with a recommendation to accept or reject the design and reasons why. A simplified 2-D analysis should be done first to identify the most critical locations. A 3-D analysis can then focus on the critical spots and verify critical values.

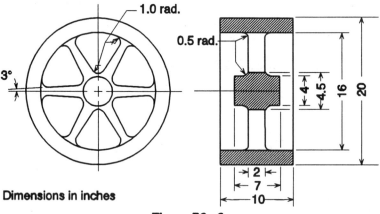

Figure P6 - 9.

References

6.1   Zienkiewicz, O. C., and Taylor, R. L., *The Finite Element Method, Volume 1, Basic Formulation and Linear Problems,* Fourth Edition, McGraw-Hill Book Company, London, 1989.

6.2   Cook, R. D., Malkus, D. S., and Plesha, M. E., *Concepts and Applications of Finite Element Analysis,* Third Edition, John Wiley and Sons, New York, 1989.

6.3   Irons, B., "Quadrature Rules for Brick Based Finite Elements," Int. J. Numerical Methods in Engineering, Vol. 3, No. 2, 1971, pp. 293-294.

CHAPTER 7

AXISYMMETRIC SOLIDS

There are many occasions in engineering where the three-dimensional structure involved will have an axisymmetric geometry with axisymmetric loads. As described in Chapter 2, the elastic structural solution for these problems simplifies a great deal, and in essence, it becomes a two-dimensional problem. While many of the modeling considerations will be the same as the 2-D plane problem, there is an additional strain and stress component included, and the application of loads to the model is different from the plane 2-D case.

In this chapter, we will only deal with the case of torsionless axisymmetry and address the torsion problem in the following chapter. Examples of the type of analysis problems solved with an axisymmetric case include centrifugal loading of a disc of constant cross section rotating about its axis of symmetry, a circular cylinder with internal and external pressure loading, or a thick wall sphere with pressure or nonuniform temperature loading. Although this axisymmetric condition is a simplified 3-D elasticity problem, getting exact closed form solutions for the general axisymmetric shape is practically impossible.

The geometric definition commonly employs a cylindrical coordinate system using the z axis as the axis of symmetry as illustrated in Figure 7-1. The cross section geometry lies in the r-z plane. This cross section shape then revolves about the z axis to form the three-dimensional solid. It was shown in Chapter 2 that the displacement is given by two components in the r-z plane, and the six 3-D strain and stress components reduce to four. They are the 2-D components in the r-z plane along with the normal strain and stress components in the hoop or θ direction.

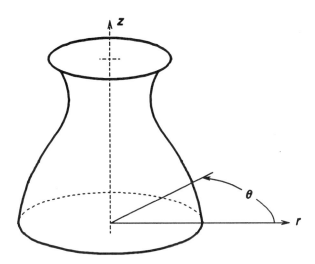

Figure 7-1. An Axisymmetric Solid in a Cylindrical Coordinate System

7.1   Element Formulation

In torsionless axisymmetric problems the displacement field consists only of the two displacement components of any material point located in the r-z plane. These correspond to the u and w components of the three-dimensional formulation. As in the 2-D plane problems, we may formulate the axisymmetric element either in a triangular or quadrilateral shape. Because of the convergence problems associated with linear triangle elements in practical analyses, we will only consider the quadrilateral. The formulation changes necessary to produce an axisymmetric element from a plane element follow.

An axisymmetric quadrilateral element appears in Figure 7-2. The node points 1, 2, 3, and 4 are actually node circles indicated by the dashed circles passing through these points in the figure. The nodes may only displace in the r direction with an increase in the circle radius or the z direction with a constant circle radius. The element formulation is very similar to the 2-D plane element so this discussion will focus on the differences and additions included for the axisymmetric conditions.

In a square element the displacement components may be approximated by the same set of polynomials used in the 2-D plane element which in r,z coordinates become

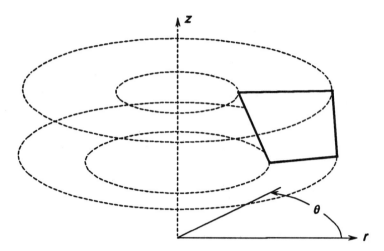

Figure 7-2. An Axisymmetric Quadrilateral Element

$$u = a_1 + a_2 r + a_3 z + a_4 rz$$
$$w = a_5 + a_6 r + a_7 z + a_8 rz \ . \tag{7.1}$$

Use of the strain-displacement relations from Chapter 2 show that the strain component approximations are

$$\epsilon_r = \frac{\partial u}{\partial r} = a_2 + a_4 z$$

$$\epsilon_\theta = \frac{u}{r} = a_2 + \frac{(a_1 + a_3 z)}{r} + a_4 z$$

$$\epsilon_z = \frac{\partial w}{\partial z} = a_7 + a_8 r \tag{7.2}$$

$$\gamma_{rz} = \frac{\partial u}{\partial z} + \frac{\partial w}{\partial r} = a_3 + a_4 r + a_6 + a_8 z \ .$$

These show the same type of variation exhibited by the 2-D plane element for the components in the r-z plane. The additional tangential component is a linear function of z and an inverse function of r. This shows a singularity in the tangential component as the material point approaches the axis of revolution. However, the theory of elasticity [7.1] shows that the tangential component becomes equal to the radial component at the axis. As in the 2-D plane case, the assumed displacement function automatically satisfies compatibility conditions.

Make the same transformation of the square element into an isoparametric quadrilateral as in the 2-D plane element and use Lagrangian interpolation.

$$u = N_1u_1 + N_2u_2 + N_3u_3 + N_4u_4$$
$$w = N_1w_1 + N_2w_2 + N_3w_3 + N_4w_4$$

(7.3)

where the interpolation formulas, N_i, are given by

$$N_1 = \frac{1}{4}(1-\xi)(1-\eta)$$

$$N_2 = \frac{1}{4}(1+\xi)(1-\eta)$$

$$N_3 = \frac{1}{4}(1+\xi)(1+\eta)$$

$$N_4 = \frac{1}{4}(1-\xi)(1+\eta) \ .$$

(7.4)

Now use the same interpolation formulas to interpolate for the coordinates of any given point r,z in an element from its node coordinates.

$$r = N_1r_1 + N_2r_2 + N_3r_3 + N_4r_4$$
$$z = N_1z_1 + N_2z_2 + N_3z_3 + N_4z_4$$

(7.5)

The element's [B] matrix is then

$$[B] = [\partial][N]$$

(7.6)

where the partial derivative operator, $[\partial]$, is given in equation (7.7) and the [B] matrix is given in equation (7.8).

$$[\partial] = \begin{bmatrix} \dfrac{\partial}{\partial r} & 0 \\[2mm] \dfrac{1}{r} & 0 \\[2mm] 0 & \dfrac{\partial}{\partial z} \\[2mm] \dfrac{\partial}{\partial z} & \dfrac{\partial}{\partial r} \end{bmatrix}$$

(7.7)

$$[B] = \begin{bmatrix} \dfrac{\partial N_1}{\partial r} & 0 & \dfrac{\partial N_2}{\partial r} & 0 & \dfrac{\partial N_3}{\partial r} & 0 & \dfrac{\partial N_4}{\partial r} & 0 \\[2em] \dfrac{N_1}{r} & 0 & \dfrac{N_2}{r} & 0 & \dfrac{N_3}{r} & 0 & \dfrac{N_4}{r} & 0 \\[2em] 0 & \dfrac{\partial N_1}{\partial z} & 0 & \dfrac{\partial N_2}{\partial z} & 0 & \dfrac{\partial N_3}{\partial z} & 0 & \dfrac{\partial N_4}{\partial z} \\[2em] \dfrac{\partial N_1}{\partial z} & \dfrac{\partial N_1}{\partial r} & \dfrac{\partial N_2}{\partial z} & \dfrac{\partial N_2}{\partial r} & \dfrac{\partial N_3}{\partial z} & \dfrac{\partial N_3}{\partial r} & \dfrac{\partial N_4}{\partial z} & \dfrac{\partial N_4}{\partial r} \end{bmatrix} \qquad (7.8)$$

The derivatives of interpolation functions shown above must be taken with respect to the natural coordinates ξ and η and then transformed to the r,z coordinates by use of the inverse of the Jacobian as shown in Chapter 5. Also, the denominator value of r above must be interpolated from the nodal values of r as given in equation (7.5).

The element stiffness matrix takes on the form given by the principle of virtual work in Chapter 5 and is repeated in equation (7.9).

$$[k] = \int_V [B]^T [E][B] dV \qquad (7.9)$$

The expression for the differential volume in the integral in cylindrical coordinates is

$$dV = r dr d\theta \, dz \, . \qquad (7.10)$$

Since there is no variation of any variable with respect to θ, then $d\theta$ integrates to 2π, and $(\det[J])d\xi d\eta$ replaces $drdz$, so the integral for the element stiffness matrix becomes

$$[k] = 2\pi \int_{-1}^{+1} \int_{-1}^{+1} [B]^T [E][B] r (\det[J]) d\xi \, d\eta \, . \qquad (7.11)$$

The coordinate r is again given by the interpolation formula given in equation (7.5).

The complexity of the integrand again demands numerical integration using Gauss quadrature. If we take the volume integral around the total circumference, the applied loads must also integrate around the total circumference to determine the nodal loads. Many programs will avoid the superfluous multiplication by 2π in each case by dividing it from both sides of the equation [7.2]. In this case the stiffness matrix becomes the stiffness for a one radian segment of the structure and the nodal loads are the total loads applied to a one radian segment of the structure. The user should check the program user's manual to determine what convention a given program uses.

Formulations for a higher order parabolic triangle or a parabolic quadrilateral element that incorporate midside nodes modify the 2-D plane formulations in the same ways. Basically, the same considerations apply for choosing between these different axisymmetric element formulations for a model as apply in the 2-D plane case.

7.2   The Finite Element Model

Most of what was learned studying 2-D model planning also applies to the axisymmetric model. For instance, the element type of choice is typically the linear quadrilateral element to begin the model studies. The mesh area will cover the cross section geometry of the axisymmetric solid defined in the r-z plane using only positive values of radius, r. The mesh subdivision plan should then properly account for the expected variations in displacement, strain, and stress throughout the area.

The fact that the 3-D solid has an axis of symmetry makes a dramatic reduction in the cost of this analysis as opposed to a fully 3-D analysis. Further, exploiting any symmetry conditions that exist in the r-z plane reduces it more. For example, the rotating wheel shown in Figure 7-3. has geometric symmetry about the r axis. Another example is the thick-wall sphere with spherically symmetric loads in Figure 7-4. There are any number of lines of symmetry in the r-z plane that come from the origin. So any wedge section of axisymmetric elements as drawn in Figure 7-5 may make a model.

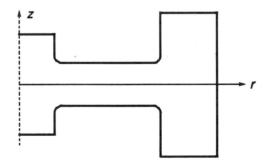

Figure 7-3. An Axisymmetric Rotating Wheel

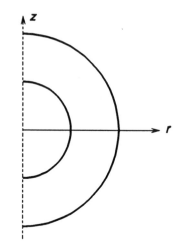

Figure 7-4. A Thick-Wall Sphere

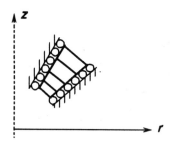

Figure 7-5. A Spherical Wedge Section

There is only one potential rigid body motion of the axisymmetric structure. It is a translation along the z axis since no other free translation or rotation can occur due to the formulation. Therefore in any model there must be at least one node restrained against z displacement. In most cases any symmetries existing in the r-z plane eliminate this mode is automatically.

In all finite element models the structure loads must resolve to node point locations. In this case the formulation has used two translation displacement components as degrees-of-freedom, therefore there are two nodal force components in the r-z plane. Do not forget, however, that in the axisymmetric model the node represents a circumferential line rather than just a point, and therefore the loads applied are line loads uniformly distributed around the node circle. If the program integrates on a one-radian segment rather than the complete circumference, then the input is the total force acting on a one-radian segment of the node circle.

In the case of pressure loading on a linear, corner-noded element face, the total force equals the pressure times the face area of a one-radian segment of the node circles. The total force is then split between the two nodes on the face.

Body forces such as the centrifugal force generated by an angular rotation about the axis-of-symmetry calculates similarly. The radial force on a one-radian volume of the element cross section divides equally to the four nodes. In this case the centrifugal force is proportional to the radius within the element, and the most proper calculation is to integrate over the element area. However, in most cases with the element subdivision fine enough the value is sufficiently accurate based upon the average or centroidal radius of the element [7.3].

In higher-order elements, the distributions to nodes must be based upon the interpolation formulas. The user must then trust the program to make this distribution properly, and it does not usually match common sense.

7.3   Computer Input Assistance

The computer input of the axisymmetric model follows the same procedures used for 2-D models with some exceptions discussed below. All the radial coordinates of points must be positive, and the axis of symmetry must pass through the origin of the global coordinate system. In most programs the choice of coordinate axis which defines the axis of symmetry is already made in the program. Many of these will use the y axis of the global cartesian coordinate system with the cross section definition in the x-y plane. Then the r coordinate is parallel to x and the z coordinate lies on the y axis. Some programs will allow the user to choose which coordinate is to be the axis of symmetry. A few programs use the z global axis for the axis of symmetry and then the model must lie in the x-z plane.

Even though the element formulation uses definition in a cylindrical coordinate system, this does not require the use of a cylindrical coordinate system in any program's preprocessor, since the model cross section lies in a cartesian plane. The various preprocessor coordinate systems are useful for defining the object geometry.

Selecting the axisymmetric element from a program's element library may be direct or may simply be an option on the 2-D plane element menu. Since the formulation is so near the 2-D plane formulation, only minor revisions of code will change the plane element to an axisymmetric element. For this reason, the program developer may choose to list the axisymmetric element as a plane element option.

We may also be able to solve problems of unsymmetric loading on axisymmetric objects if the load distribution can resolve to a Fourier series of components. In this case, once we fit the load distribution by the

Fourier series, then each Fourier amplitude coefficient defines a load case. All cases are run, and superposition of the results from all the load cases finds the final solution. This comes about by assuming that all three displacement components exist for the axisymmetric 3-D object. Then the total displacement components are a combination of the amplitude at $\theta = 0.0$ times sines and cosines of $n\theta$, where n is the integer number of cycles in the Fourier components. This approach can be useful in some cases, and although it is not widely available in commercial programs, there are some which include this capability. However, unless we can fit the load distribution with a relatively few Fourier coefficients, then the more practical approach may be to turn to a fully three-dimensional solution.

7.4   The Analysis Step

Review the material in Chapter 5 on 2-D plane analysis and apply the same concerns to axisymmetric analysis. Numerical errors caused by element distortion, compatibility violations and ill-conditioning are also of concern here.

In addition to the concerns expressed in the 2-D analysis chapter, there is a special concern with axisymmetric structures. It is that the user has defined the solid correctly. An incorrect definition will occur whenever the defined cross section does not lie in the proper plane selected by the program. In 2-D plane models the location and orientation of the model in the coordinate system is of no importance, but in axisymmetric solids both the location and orientation are critical. For example, if the program expects the model definition to use the y axis of its global cartesian coordinate system as the axis of symmetry and the user defines the body as if the x axis is the axis of symmetry, then the volume is incorrect. Since the graphical display only shows the cross section and not the solid body, the error is difficult to catch. Similarly, if the program expects the cross section definition in its cartesian x-z plane, and the user defines the cross section in the x-y plane, then the swept volume produced by revolution about the z axis will have zero thickness.

7.5   Output Processing and Evaluation

Once again the guidelines established in the 2-D chapter apply to the results from the axisymmetric analysis. Contour plots develop in the same manner and with the same potential to show misleading results. The results must converge to acceptable accuracy.

Remember, there is the additional hoop stress component for display. Usually this component is the largest of the stress components. The three-dimensional nature of the stress state makes evaluation based on the Von Mises effective stress even more critical for an accurate representation of the failure potential of a structure.

7.6   Case Studies

We will explore three different cases in axisymmetric geometries. The first case will be a capped thick-wall cylinder with internal pressure. The second is a thick-wall sphere with internal pressure, and the third is a circular plate with center load.

The capped cylinder with internal pressure is in Figure 7-6. The cylinder is 6 in. outside diameter, 4 in. inside diameter, and 8 in. long. The end cap plates are 1 in. thick and there is a 0.5 in. radius fillet on the inside corner.

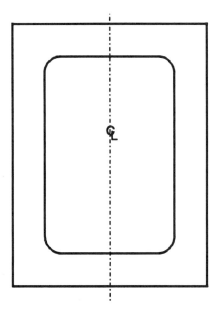

Figure 7-6. A Capped Cylinder Pressure Vessel

An axisymmetric element mesh is shown in Figure 7-7 which is symmetric about the cylinder mid-plane. The boundary conditions enforce vertical displacement restraint of nodes on the mid-plane and radial displacement restraint of nodes on the centerline or axis of revolution. There is an applied uniform internal pressure of 1000 psi symbolized by the double line along the inner surface.

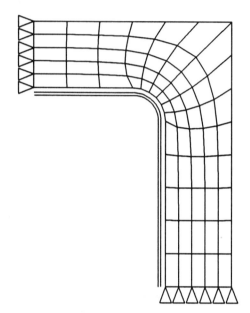

Figure 7-7. A Symmetric Model of the Cylinder Pressure Vessel

After running the solution, the deformed shape geometry is given in Figure 7-8. The plot shows the correct boundary conditions enforcement, and it shows a reasonable deformation pattern. The cylinder should expand radially and the cap should bulge at the center.

Checking the stress results begins with the X-stress or radial stress since the axis of revolution is the global *y* coordinate axis. The radial stress contour plot is given in Figure 7-9. The radial stress on the inside diameter of the cylinder must be equal to the pressure at -1000 psi. This should show contour level 2 close to the inner surface so in this case the finite element result underpredicts the actual value. The radial stress on the outside diameter is zero, which is between contours 3 and 4. This is in reasonable agreement with the contour levels plotted near the outside diameter surface. There is some concentration of radial stress near the top of the inside fillet reaching a value of a little more than 3000 psi in tension.

Continuing with the Y-stress or axial stress in the cylinder vessel, its contour plot is in Figure 7-10. The inner surface of the cap has an applied normal stress equal to the pressure, and the agreement is reasonable across the cap inner surface. The outside surface value of zero is also reasonably accurate. These results do show the need for further model refinement to improve the accuracy. The axial stress in a long capped cylinder pipe is equal to the load due to the pressure on the bore in the

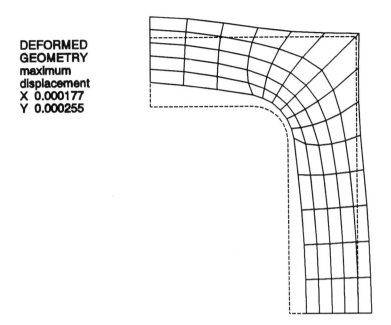

DEFORMED
GEOMETRY
maximum
displacement
X 0.000177
Y 0.000255

Figure 7-8. The Deformed Geometry of the Cylinder Pressure Vessel

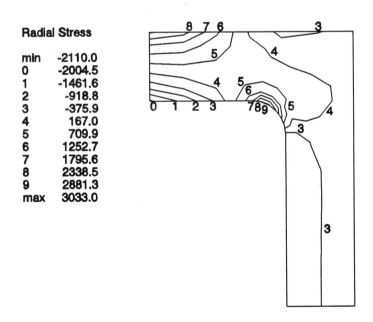

Radial Stress

min	-2110.0
0	-2004.5
1	-1461.6
2	-918.8
3	-375.9
4	167.0
5	709.9
6	1252.7
7	1795.6
8	2338.5
9	2881.3
max	3033.0

Figure 7-9. The Radial Stress in the Cylinder Pressure Vessel

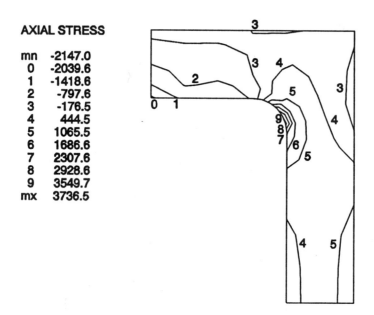

AXIAL STRESS

mn	-2147.0
0	-2039.6
1	-1418.6
2	-797.6
3	-176.5
4	444.5
5	1065.5
6	1686.6
7	2307.6
8	2928.6
9	3549.7
mx	3736.5

Figure 7-10. The Axial Stress in the Cylinder Pressure Vessel

axial direction divided by the annular cylinder area. In this case the value is 800 psi. This agreement is good on average, but the cylinder is not very long and end effects are causing some of the variation through the wall.

The third normal stress is the hoop or tangential stress labeled hoop stress in Figure 7-11. The hoop stress in a cylinder without end effects would be 2600 psi on the inside wall. This agrees well with the contour level near the mid-plane.

To predict yielding, or failure, of the vessel we must select a good yield criterion. The maximum normal stress failure theory has some serious shortcomings and we should not use it. The best criterion is the Von Mises distortion energy theory. A plot of the Von Mises equivalent stress is given in Figure 7-12. By the Von Mises distortion energy yield criterion this vessel would be at the onset of yielding if the pressure increases until the maximum equivalent stress reaches the yield strength of the material. Of course, there are sections of this model needing further refinement before we make any final judgment.

Now, turn to the second case, a uniform thick-wall sphere with internal pressure. There are any number of planes of symmetry which pass through the center of the sphere since it is point symmetric about its center. Therefore a wedge of axisymmetric elements can serve as the model. A sphere with an inside radius of 1 in. and outside radius of 2 in. fits the 15 degree segment shown in Figure 7-13. The model has three radial rows of almost square elements. Only one row is necessary, but

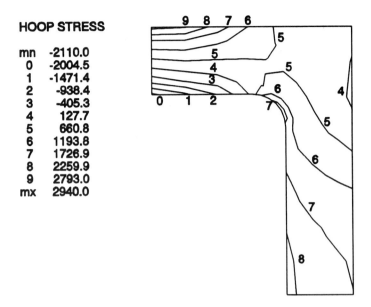

HOOP STRESS

mn	-2110.0
0	-2004.5
1	-1471.4
2	-938.4
3	-405.3
4	127.7
5	660.8
6	1193.8
7	1726.9
8	2259.9
9	2793.0
mx	2940.0

Figure 7-11. The Hoop Stress in the Cylinder Pressure Vessel

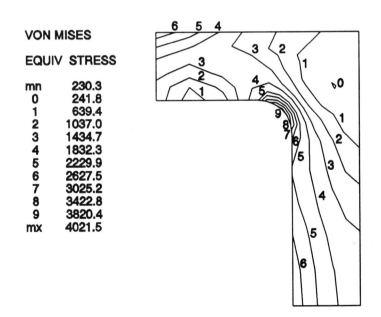

VON MISES

EQUIV STRESS

mn	230.3
0	241.8
1	639.4
2	1037.0
3	1434.7
4	1832.3
5	2229.9
6	2627.5
7	3025.2
8	3422.8
9	3820.4
mx	4021.5

Figure 7-12. The Von Mises Equivalent Stress in the Cylinder Vessel

three are used here to make the graphics a little more clear. Inclined boundary conditions make the nodes move radially in responding to the nodal loads due to pressure.

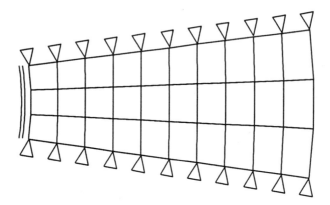

Figure 7-13. Model of a Thick-Wall Spherical Pressure Vessel

The deformed shape graphic shows the nodes moving radially, so examine the stress plots next. The X-stress component along the horizontal centerline of the model is the same as the radial stress and is in Figure 7-14. With an internal pressure of 1000 psi, the boundary values should be -1000 psi inside and zero outside. The results are in qualitative agreement, but the accuracy is not high. Notice that the contour interval is gradually decreasing toward the inside surface until the large jump from the final contour to the wall.

RADIAL STRESS

min	-744.5
0	-707.3
1	-630.7
2	-554.1
3	-477.6
4	-401.0
5	-324.4
6	-247.8
7	-171.2
8	-94.7
9	-18.1
max	-17.2

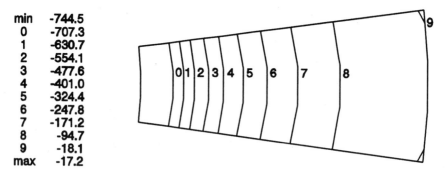

Figure 7-14. Radial Stress in the Thick-Wall Spherical Pressure Vessel

This demonstrates a quirk of the isoparametric quadrilateral element formulation. In this situation where the Y-stress component does not vary in the *y* direction it forces the X-stress variation in the *x* direction to have an inverse slope to that expected (which is to decrease going away from the wall). This does not fit the actual variation, but at nodes away from the inner wall nodal averaging tends to cancel this error. However, the value at the inside node drops off as seen in the graphic display, and underpredicts the value of radial stress.

Checking the Y-stress, the vertical hoop stress on the horizontal centerline, an increasing gradient toward the inside surface is seen in Figure 7-15. The exact value for the hoop stress in this thick sphere is 714 psi. This finite element model overestimates the hoop stress by about 13 percent. Almost the same value exists for the tangential direction hoop stress in Figure 7-16. These values are theoretically equal.

AXIAL STRESS

mn	202.0
0	212.1
1	274.2
2	336.4
3	398.5
4	460.7
5	522.8
6	584.9
7	647.1
8	709.2
9	771.4
mx	811.9

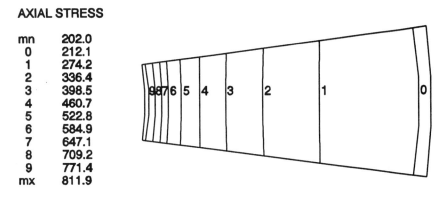

Figure 7-15. Vertical Hoop Stress in the Spherical Vessel

The Von Mises equivalent stress is in Figure 7-17. You should note the significantly higher value relative to the three normal stress components that are principal stresses. If we use the highest of the normal stresses in prediction of yield of the vessel based on the maximum normal stress failure theory, we would be in error by 48 percent!

The third case is a circular plate with a center load, and the model uses axisymmetric solid elements rather than plate elements. This plate is relatively thick even though it fits the thin plate class. However, if it was much thinner it would be difficult to model reliably with axisymmetric solid elements. In that case, plate elements would be a better choice. The plate has a 10 in. radius with a 1 in. thickness. The axisymmetric model is in Figure 7-18. The model requires a reasonable number of elements through the thickness to fit the bending stress distribution and therefore requires many elements along the radius to keep the element aspect ratio within reasonable bounds.

HOOP STRESS

min	204.2
0	214.4
1	276.8
2	339.3
3	401.7
4	464.2
5	526.6
6	589.0
7	651.5
8	713.9
9	776.3
max	817.2

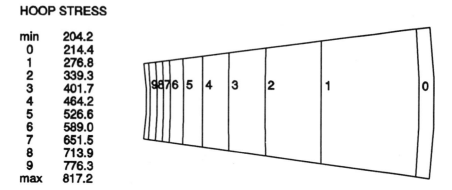

Figure 7-16. Tangential Hoop Stress in the Spherical Vessel

VON MISES
EQUIV. STRESS

min	224.6
0	235.8
1	374.0
2	512.2
3	650.5
4	788.7
5	926.9
6	1065.2
7	1203.4
8	1341.6
9	1479.8
max	1557.7

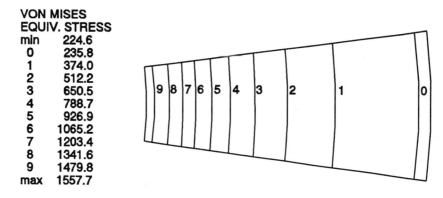

Figure 7-17. Von Mises Equivalent Stress in the Spherical Vessel

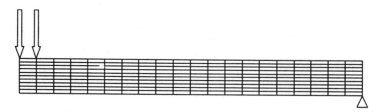

Figure 7-18. Axisymmetric Model of a Circular Plate With Center Load

The boundary conditions show a simple support around the outer edge of the plate which is at the right edge of the model. Radial restraints apply to the centerline of the plate which is at the left edge of the model. These radial restraints are usually automatic in most commercial codes, but it doesn't hurt to be sure. The center load spreads over a 0.5 in. radius circle at the center to avoid the stress singularity of a concentrated force.

The deformed shape of the plate is given in Figure 7-19. The shape is quite reasonable, and it fits the boundary conditions.

DEFORMED
GEOMETRY
maximum
displacement
X 0.000113
Y -0.001890

Figure 7-19. Deformed Shape of the Circular Plate With Center Load

In Figure 7-20 the radial or X-stress is shown, and a zoomed view of the center region appears in Figure 7-21. The boundary value of radial stress at the outer edge with a simply supported boundary is zero, and the plot agrees. At the center the maximum stress is 3512 psi compression on the top surface. Classical plate theory gives a value of 2330 psi, but it does not account for the local effect of the applied load as does the finite element model. It does agree very well with the tensile stress on the bottom surface. Examination of the contours at the center show a change in slope of the contour lines inside the core column of elements. Further refinement in this area with more radial subdivision will improve the results.

RADIAL STRESS

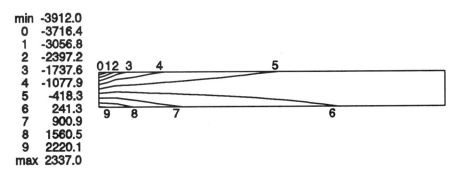

min	-3912.0
0	-3716.4
1	-3056.8
2	-2397.2
3	-1737.6
4	-1077.9
5	-418.3
6	241.3
7	900.9
8	1560.5
9	2220.1
max	2337.0

Figure 7-20. Radial Stress Contours in the Circular Plate

The tangential stress contours in Figure 7-22 show a similar agreement with the plate theory result. So overall this model gives good results, but some additional refinement would improve the agreement.

RADIAL STRESS

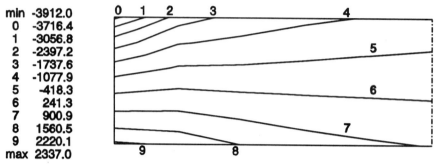

min	-3912.0
0	-3716.4
1	-3056.8
2	-2397.2
3	-1737.6
4	-1077.9
5	-418.3
6	241.3
7	900.9
8	1560.5
9	2220.1
max	2337.0

Figure 7-21. Zoom View of Radial Stress at the Plate Center

HOOP STRESS

min	-3912.0
0	-3716.4
1	-3056.8
2	-2397.2
3	-1737.6
4	-1077.9
5	-418.3
6	241.3
7	900.9
8	1560.5
9	2220.1
max	2337.0

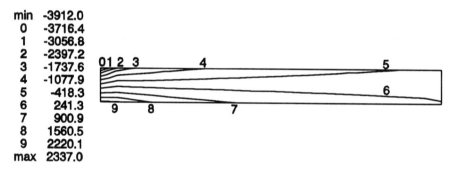

Figure 7-22. Tangential Stress Contours in the Circular Plate

Finally, Figure 7-23 presents the Von Mises equivalent stress contour plot zoomed in on the center region. Plate theory would demand that the zero contour intersect the centerline since the stresses are due to flexure and must be zero at the mid-plane of the plate. However, the finite element model includes the vertical compressive stress due to loading which decays to zero at the bottom surface.

These three cases illustrate the use of axisymmetric elements, and each case needs additional refinement to get acceptable accuracy of results. The analyst should never stop with the results from creation and execution of one finite element model.

VON MISES
EQUIV. STRESS

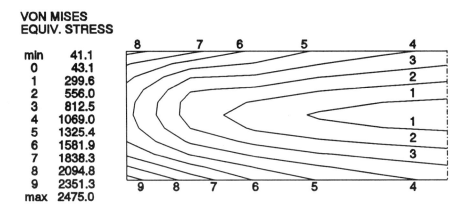

min	41.1
0	43.1
1	299.6
2	556.0
3	812.5
4	1069.0
5	1325.4
6	1581.9
7	1838.3
8	2094.8
9	2351.3
max	2475.0

Figure 7-23. Zoom View of Von Mises Stress at the Plate Center

Problems

7.1 Figure P7-1 shows a stepped shaft with a fillet radius under axial loads. Fatigue analysis for reversed axial loading requires an accurate stress concentration factor to apply to the average axial stress. Determine the stress concentration factor for the geometry shown and compare with published results. Do the values of modulus of elasticity and Poisson's ratio affect the stress concentration factor?

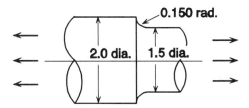

Figure P7 - 1.

7.2 The steel shaft section on the left in Figure P7-2 has a grinding relief groove machined into the diameter thus reducing the net stress area for the axial load. The section on the right puts the relief groove into the step face. Compare the stress concentration factors for the two cases if the nominal stress is calculated on the 2 in diameter area. How do these compare with any published data on the same or similar geometries?

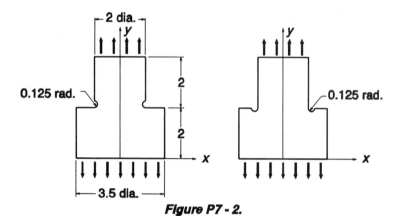

Figure P7 - 2.

7.3   A high pressure steel thick-wall cylinder design for containment of
a hot liquid is proposed as illustrated in Figure P7-3. It has a two
layer construction with cooling grooves machined into the outer
surface of the inside layer. The outside diameter is 8 in., and the
inside diameter is 5 in., with an interface diameter of 6.5 in. The
cooling grooves are semi-circular with a 0.2 in. radius and spaced at
1 inch center-to-center intervals along the cylinder axis. Compare
the structural capacity of this design for internal pressure loading
with a solid wall cylinder of the same dimensions. It is a closed
cylinder so there is axial load to apply that is proportional to the
magnitude of the pressure.

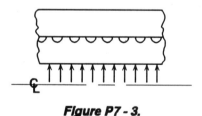

Figure P7 - 3.

7.4   A steel cylindrical pressure vessel with flat plate end caps is shown
in Figure P7-4 with a vertical axis of symmetry. Addition of
thickened sections help to reduce stress buildup in the corners.
Analyze the design and identify the most critically stressed regions.
Inside sharp corners produce infinite stress concentration zones, so
do not refine the mesh excessively at those spots or else introduce
a small fillet radius there.

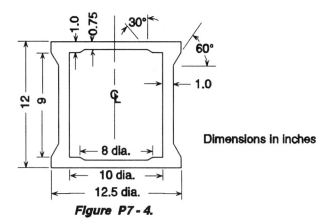

Figure P7 - 4.

7.5 A steel pressure vessel cross section is shown in Figure P7-5. The axis of symmetry is vertical, and the top plane is a plane of symmetry for the vessel. What steel should we use if the vessel is to operate safely at 5000 psi pressure?

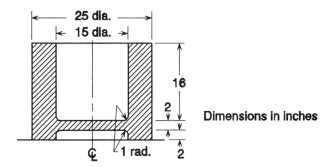

Figure P7 - 5.

7.6 A high-pressure (30,000 psi) steel extrusion barrel bolts to a relatively rigid die through the flange illustrated in Figure P7-6. The conventional design approach uses the stress equations for thick-wall cylinders on each of the two cylinder sections separately and applies a high stress concentration factor to the hoop stress on the bolt circle in the flange section. This approach ignores any stress concentration effect at the ¼-in fillet radius. Evaluate this design approach using a finite element analysis. Use an axisymmetric analysis, but consider if it warrants a 3-D analysis. You may assume that the pressure seals near the 4-in. ID and that the cylinder wall has an axial load due to pressure acting on the right end and reacted at the bolt circle.

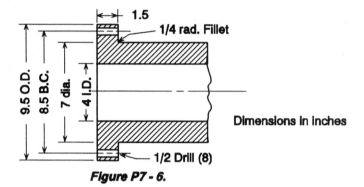

Figure P7 - 6.

7.7   The Belleville spring in Figure P7-7 is made from 0.050-in. thick steel. These springs typically have a very nonlinear stiffness if the vertical compression displacement is a significant value compared with the spring height. Apply a line load on the upper edge and vertical restraint on the lower edge. Determine the initial stiffness of this spring. How much does it change if you also restrain the lower edge against horizontal motion? How might you determine if there was significant nonlinearity in the stiffness for the range of applied loads in any given application?

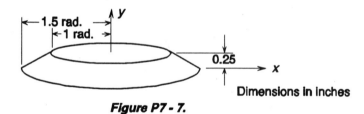

Figure P7 - 7.

7.8   A thick hub aluminum rotor such as that sketched in Figure P7-8 is a common component in rotation machinery. The initial design approach is to use stress equations for a rotating thin disk of constant thickness. How well are the stresses that develop in the design shown predicted by the equations for a thin rotating disk? Remember that the material specific weight or mass density (depending on individual program requirements) must be input in the correct system of units to apply the centrifugal loading.

7.9   The spur gear in Figure P7-9 must operate at speeds up to 6000 rpm. The gear originally had straight sides with a constant thickness of 12 mm. In order to reduce the rotational inertia and

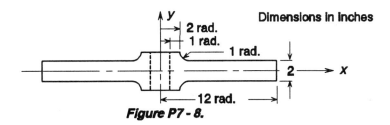

Figure P7 - 8.

improve the angular acceleration response, the sides were machined down to 6 mm thick between the hub and rim. The gear material is UNS G10150 cold drawn steel. Has the structural integrity of the gear been reduced? In modeling by an axisymmetric analysis, account for the material loss in the space between teeth. If there were a keyway on the bore, estimate the potential for stress fracture in both cases.

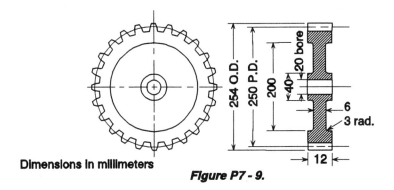

Dimensions in millimeters

Figure P7 - 9.

7.10 Analyze the bolted, capped end of the steel cylindrical pressure vessel in Figure P7-10. There are ten 1-in. diameter bolts preloaded to produce gasket sealing with an average gasket compressive stress of 10.8 kpsi. The operating pressure is 3500 psi. The gasket is copper-clad asbestos with a modulus of elasticity of $E = 13.4$ Mpsi. In an axisymmetric analysis use a truss element to represent the bolts, and connect it to the cap and vessel where the bolt head and nut contact respectively. First apply the bolt preload by "tightening". Do this by way of initial strain in the element or an equivalent thermal shrinkage until the average compressive stress in the gasket is 10.8 kpsi. Then apply the internal pressure load. To maintain the gasket seal, the average gasket compressive stress should be at least 2.5 times the internal pressure. Is this requirement satisfied here? Are there any critical stress areas in this design if a normal carbon steel is the material of construction? Should a 3-D analysis be performed?

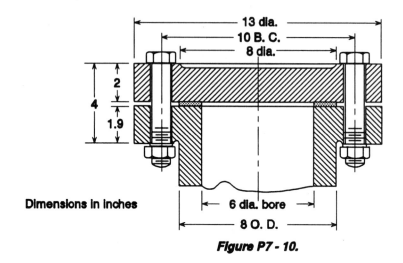

Dimensions in inches

Figure P7 - 10.

References

7.1   Timoshenko, S. and Goodier, J. N., *Theory of Elasticity*, McGraw-Hill, New York, 1951.

7.2   Cook, R. D., Malkus, D. S., and Plesha, M. E., *Concepts and Applications of Finite Element Analysis*, Third Edition, John Wiley and Sons, New York, 1989.

7.3   Zienkiewicz, O. C., and Taylor, R. L., *The Finite Element Method, Volume 1, Basic Formulation and Linear Problems*, Fourth Edition, McGraw-Hill Book Company, London, 1989.

CHAPTER 8

TORSION OF PRISMATIC AND AXISYMMETRIC SOLIDS

Loading of a general shaped three-dimensional object by torsional loads obviously requires a 3-D analysis. However, two special cases of torsion loading of prismatic or axisymmetric geometries are very common. A prismatic geometry consists of a two-dimensional cross section shape which is constant along the entire length of the member. Torsion loads apply at stations along the length of a member and cause twisting about the axis running the length of the member. A square cross section prismatic bar was shown in Figure 2-9 of Chapter 2. An axisymmetric geometry forms by revolution of a two-dimensional cross section about an axis. Torsion loads may act at stations along the axis of symmetry. An axisymmetric bar in torsion was shown in Figure 2-10 of Chapter 2.

Solutions for these two cases are much simpler than the general three-dimensional case. Special 2-D formulations result for each of these problems [8.1]. Of course, the case of torsion loading on a solid circular cylinder or tube has a very simple exact solution as presented in any introductory mechanics of materials text. As discussed in Chapter 2, solutions for these two cases derive from a procedure in theory of elasticity called the semi-inverse method and result in a single partial differential equation for each case.

The partial differential equation for torsion of a prismatic bar takes the form of the Laplace equation with the independent variable being the warping function over the area of the cross section. The partial differential equation for the axisymmetric body involves a defined twisting function. We can get solutions for each of these equations using several different numerical methods for simple configurations. However, the finite element formulation works better for complex configurations.

8.1   Element Formulation

The elements for the two different cases formulate similarly but have different resulting forms. We will develop the element for torsion of the noncircular prismatic bar first. The discussion in Chapter 2 presented the basic assumptions involving the displacement field. There, it was shown that the displacement field was given completly by geometric position and the warping function, ψ, of the cross section. Expressions for strain and stress that result from this displacement field were also given. Once we find the warping function over the cross section area, we may compute the strains.

The finite element formulation begins by assuming a distribution of the warping function over the area of an element in much the same manner used for displacement components in other elements. The assumption is

$$\psi = [N]\{\psi_i\} \tag{8.1}$$

where, $[N]$ is the interpolation function matrix for the chosen element geometry, and ψ_i are the node values of the warping function [8.1].

Recall the elasticity relations from Chapter 2 for the displacements in equation (8.2), strains in equation (8.3), and stresses in equation (8.4).

$$u = -\theta z y$$
$$v = \theta z x \tag{8.2}$$
$$w = \theta \psi$$

$$\gamma_{zx} = \theta \left(\frac{\partial \psi}{\partial x} - y \right)$$
$$\gamma_{zy} = \theta \left(\frac{\partial \psi}{\partial y} + x \right) \tag{8.3}$$

$$\begin{Bmatrix} \tau_{zx} \\ \tau_{zy} \end{Bmatrix} = \begin{bmatrix} G & 0 \\ 0 & G \end{bmatrix} \begin{Bmatrix} \gamma_{zx} \\ \gamma_{zy} \end{Bmatrix} \tag{8.4}$$

Application of the principle of virtual work with this assumption forms the basis of the finite element formulation. Now for any given set of small virtual displacements the internal virtual strain energy, δU_e, becomes

$$\delta U_e = \int_V \{\delta \gamma\}^T \{\tau\} dV \qquad (8.5)$$

where, $\{\delta \gamma\}$ are the virtual strain components produced by the small virtual displacements, $\{\tau\}$ are the stress components in the differential material volume at equilibrium, and dV indicates the differential volume element within the finite element.

Making the substitutions for strain and stress components from the relations above produces

$$\delta U_e = \int_V \delta \theta \left\{ \begin{array}{c} \dfrac{\partial \delta \psi}{\partial x} - y \\[2mm] \dfrac{\partial \delta \psi}{\partial y} + x \end{array} \right\}^T \begin{bmatrix} G & 0 \\ 0 & G \end{bmatrix} \theta \left\{ \begin{array}{c} \dfrac{\partial \psi}{\partial x} - y \\[2mm] \dfrac{\partial \psi}{\partial y} + x \end{array} \right\} dV \qquad (8.6)$$

where, $\delta \theta$ and $\delta \psi$ are small virtual displacements from the equilibrium configuration, and θ and ψ are the actual displacements of the material from the unloaded to the equilibrium position under load. Integrating in the axial, z, direction and multiplying terms results in equation (8.7).

$$\delta U_e = (\delta \theta) \theta \, Gz \int_A \left(\frac{\partial \delta \psi}{\partial x} \frac{\partial \psi}{\partial x} - y \frac{\partial \delta \psi}{\partial x} - y \frac{\partial \psi}{\partial x} + y^2 + \right.$$
$$\left. \frac{\partial \delta \psi}{\partial y} \frac{\partial \psi}{\partial y} + x \frac{\partial \delta \psi}{\partial y} + x \frac{\partial \psi}{\partial y} + x^2 \right) dA \qquad (8.7)$$

The external virtual work of torsion stresses over the end area is given by equation (8.8), where, δu and δv are the virtual displacements.

$$\delta W_e = \int_A (\delta u \tau_{zx} + \delta v \tau_{zy}) dA \qquad (8.8)$$

Making the substitutions for displacement and stress components from the relations above produces equation (8.9), where, $\delta \theta$ is the virtual angle of torsion and θ is the actual angle of torsion.

$$\delta W_e = \int_A \left(-\delta \theta \, zy G \theta \left(\frac{\partial \psi}{\partial x} - y \right) + \delta \theta \, zx G \theta \left(\frac{\partial \psi}{\partial y} + x \right) \right) dA \qquad (8.9)$$

Factoring out $(\delta \theta) \theta Gz$ and setting the virtual strain energy equal to the virtual work yields

$$\int_A \left(\frac{\partial \delta \psi}{\partial x} \frac{\partial \psi}{\partial x} + \frac{\partial \delta \psi}{\partial y} \frac{\partial \psi}{\partial y} \right) dA = \int_A \left(y \frac{\partial \delta \psi}{\partial x} - x \frac{\partial \delta \psi}{\partial y} \right) dA \qquad (8.10)$$

Use the interpolation functions for ψ in the equation above by

$$\frac{\partial \delta \psi}{\partial x} = \frac{\partial [N]}{\partial x} \{\delta \psi_i\} = \{\delta \psi_i\}^T \frac{\partial [N]^T}{\partial x} \qquad (8.11)$$

with similar relations for the other derivatives. This derives the finite element equations (8.12) below.

$$\{\delta \psi_i\}^T \left[\int_A \left(\frac{\partial [N]^T}{\partial x} \frac{\partial [N]}{\partial x} + \frac{\partial [N]^T}{\partial y} \frac{\partial [N]}{\partial y} \right) dA \right] \{\psi_i\}$$

$$= \{\delta \psi_i\}^T \int_A \left(y \frac{\partial [N]^T}{\partial x} - x \frac{\partial [N]^T}{\partial y} \right) dA \qquad (8.12)$$

Since the equation must be valid for any set of virtual displacements, the virtual node values, $\{\delta \psi_i\}^T$ may be removed from each side of the equation, and the element stiffness matrix is given in equation (8.13), with node forces given by equation (8.14). The finite element equations are then given in equation (8.15).

$$[k] = \int_A \left(\frac{\partial [N]^T}{\partial x} \frac{\partial [N]}{\partial x} + \frac{\partial [N]^T}{\partial y} \frac{\partial [N]}{\partial y} \right) dA \qquad (8.13)$$

$$\{f\} = \int_A \left(y \frac{\partial [N]^T}{\partial x} - x \frac{\partial [N]^T}{\partial y} \right) dA \qquad (8.14)$$

$$[k]\{\psi\} = \{f\} \qquad (8.15)$$

At this point we may adopt either a linear or parabolic triangle or quadrilateral and its corresponding interpolation functions from previous

2-D elements to continue the element formulation. Notice that in this case there is only one solution variable component thus there is only a single degree-of-freedom per node.

Rather than continue a specific element formulation, it is sufficient at this point to note some characteristics of the elements. In a linear triangle element formulation, the stress components within the element have a portion that is constant from the derivative of the warping function, and a portion that varies linearly with distance from the torsion axis as shown in equation (8.3). In an isoparametric quadrilateral element, the stress portion due to the derivative of the warping function also allows a variation within the element field and with the variation allowed by the distance from the torsion axis should provide a better fit to the actual distribution.

Of course, it is assumed that the member is long relative to its cross section dimensions, and the distributions from the solution will be valid on cross sections away from the ends according to St. Venant's principle [8.1]. Unless the end torque distribution matches that found in the solution, the stresses at the ends will be different where the torque applies, and it becomes a 3-D problem at the ends.

Now turn to torsion of axisymmetric circular cross section members with variable diameter along the length. This is very similar to the conditions for axisymmetric analyses, except that torsional loading applies to the body of revolution. The elasticity formulation by the semi-inverse method presented in Chapter 2 assumes that only the tangential or circumferential component of displacement is nonzero [8.1].

Now approximate the twisting function, ϕ, within an element by

$$\phi = [N]\{\phi_i\} \tag{8.16}$$

where, the shape functions, $[N]$, are those appropriate for the chosen element, and $\{\phi_i\}$ are the node values of the twisting function. Through the principle of virtual work [8.2] the twisting stiffness matrix takes the form

$$[k] = G \int_A r^3 [B]^T [B] dr dz \tag{8.17}$$

where, A is the element area, and the $[B]$ matrix is shown in equation (8.18). The node point load components are found using equation (8.19), where, $[\tau]$ is the tangential shear stress as a function of r on the ends where there is torque load. The finite element equations are then shown in equation (8.20).

Again, we may chose either a linear or parabolic triangle or quadrilateral to complete the element formulation.

$$[B] = \begin{bmatrix} \dfrac{\partial[N]}{\partial r} \\[2ex] \dfrac{\partial[N]}{\partial z} \end{bmatrix} \tag{8.18}$$

$$[f] = \int_r r^2 [N]^{\mathsf{T}} [\tau] dr \tag{8.19}$$

$$[k]\{\phi\} = [f] \tag{8.20}$$

8.2   The Finite Element Model

Even though these cases involve a 3-D solid object it has now reduced to a two-dimensional analysis. Although these elements are not very complex, they are rarely available in the element libraries of commercial finite element programs. Therefore, to solve these problems you may have to use an analogous heat transfer solution, use 3-D elements, or resort to writing your own program. The finite element mesh plan for the prismatic bar torsion case will lie within the cross section geometry. The mesh plan should properly account for areas of expected rapid variations of the warping function and its corresponding shear stresses.

Symmetry conditions that exist in any of these area descriptions help reduce the model size. Warping of the cross section of a prismatic bar will result in some portions displacing in the direction of the outward normal to the cross section plane while the other portions displace in the opposite direction. The displacement direction will change across each line of symmetry of the cross section. On these lines of symmetry the warping function must be zero. For example, in torsion of a shaft with a square cross section, the model utilizes lines of symmetry passing through the center horizontally and vertically and through the corners as shown in Figure 8-1. This reduces the model to the one-eighth section shown shaded in the figure. A rectangular cross section would only have quarter symmetry. Load inputs for prismatic bar torsion were given in equation (8.14). The program must evaluate this expression on an element-by-element basis.

Though a program may not include these torsion elements, elements formulated for heat transfer can solve one of these problems. For torsion of a noncircular cross section we may use the 2-D conduction heat transfer element. The governing equation for steady state conduction heat transfer is nearly identical to that of this torsion case. You then only need to make the right property and variable correlations and interpretations to find the torsion solution. The conduction heat transfer finite element formulation

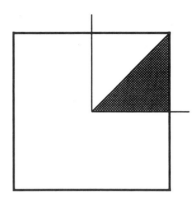

Figure 8-1. A Square Cross Section Shaft in Torsion

will be discussed in a later chapter, but some of the correspondence of properties and variables are shown in Table 8-1. Of course, this presumes you have a heat transfer program available for use.

Table 8-1.   Correspondence of Heat Transfer and Torsion Solutions

Parameter	Heat transfer	Torsion
Nodal Variable	Temperature	Warping Function
Material Property	Conductivity	Shear Modulus
Load	Heat Flux	Warping Gradient

To use the heat transfer program make the substitution of variables listed in Table 8-1. Since we factored out the shear modulus, G, and the angle of twist, θ, from equation (8.9), use a conductivity value of unity. Using a value of unity will make the conduction element stiffness matrix agree with equation (8.13). The element nodal loads in equation (8.14) are self-equilibrating on all nodes internal to the boundary. On the boundary they become equivalent to a specified heat flux normal to the boundary (the flux value will make the shear stress on the free surface equal to zero). The x component of this flux is linearly proportional to the y coordinate at any boundary point. The y component is linearly proportional to and the negative of the x coordinate of the boundary point. Multiply the x component by the cosine of the angle between the boundary normal and the x axis. Multiply the y component by the sine of the angle and sum the two values to get the normal heat flux that applies at specific boundary points. Do this for all the boundary points except those on symmetry lines having a specified warping value of zero. The solution for temperature will now correspond to the warping function for the cross section.

Using the warping function solution, we can find the shear strains from equation (8.3) for an arbitrarily selected value of θ and the shear stresses from equation (8.4). Finally, the torque resulting from the selected value of θ comes from the integral over the cross section in equation (8.21).

$$T = \int_A (-y\tau_{zx} + x\tau_{zy})dA \qquad (8.21)$$

There also is an analogy between the heat transfer solution and the stress function formulation of the torsion problem. This approach solves for the stress function over the cross section instead of the warping function. It is similar to the displacement formulation, but it is limited in application to singly connected regions (no holes in the cross section allowed), while the displacement formulation is not. We have chosen to present the displacement formulation here because of its general applicability. Most texts and programs usually present the stress function analogy.

Another solution approach is to use 3-D solid elements. The obvious model is a long bar with mesh subdivision over the cross section and along the length of the bar with torsion loads at the ends. The disadvantages are that the model becomes large and, unless we know the correct distribution of torsion loading, the solution is invalid near the ends. If we invoke Saint-Venant's principle, the solution away from the ends should be correct. However, we can construct a much more efficient 3-D model.

A displacement loaded, one-element thick 3-D model will give an efficient solution without the confusing aspects of the heat transfer analogy. The Saint-Venant torsion theory prescribes the displacement conditions acting on the model. Node point displacement components in the plane of the cross section only vary linearly with the z coordinate. The displacement component in the z direction is constant along z. Therefore, we need only one element in the z direction of the model. An example model is shown in Figure 8-2.

Assume this model is one-quarter section of a square shaft one unit in length. We need to apply displacement conditions to the model that agree with the Saint-Venant torsion theory. Set to zero the x and y displacements of all nodes lying on the x-y plane at $z = 0$. Prescribe a rotation about the z axis for all nodes on the x-y plane at $z = 1$. One way to accomplish this is to change the nodal displacement coordinate system for these nodes to a cylindrical system with its origin at the z axis. Then set all the radial displacements to zero and all the tangential displacements to values linearly proportional to radius that correspond to an arbitrary value of angle of twist per unit length, θ. Finally, set all the z displacements to zero for all nodes lying on planes of symmetry of the cross section. In this example, those will be nodes on the x-z and y-z planes. This completes the model, and it is ready for solution.

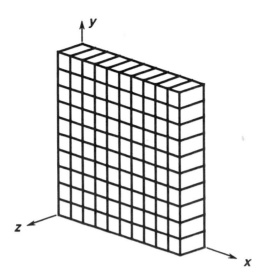

Figure 8-2. A One-Quarter Section Square Torsion Bar 3-D Model

Following solution, we can examine the plot of z displacements on the cross section to see the warping effects. Next, plots of τ_{zx} and τ_{zy} show us the shear stress distributions. We can get the torsional constant for the cross section by first finding the total reaction torque about the z axis of the nodal reaction forces for nodes on the x-y plane at $z = 0$. Then the torsional constant, J, is given in equation (8.22), where T is the reaction torque times 4 (one-quarter model), L is one unit length, θ is the section rotation angle in radians, and G is the material shear modulus.

$$J = \frac{TL}{\theta G} \tag{8.22}$$

Now we turn to analysis of torsion loading of an axisymmetric member. If the torsion element is not available, we can solve this problem using the axisymmetric element described in the previous chapter, provided the program can apply anti-symmetric loading. If it can, then solve the torsion problem by application of a circumferential load. In this solution the nodal variable will be the circumferential displacement rather than the twisting function. Also, a 3-D solid element model with prescribed displacements will solve the problem.

The mesh plan for 2-D analysis of the axisymmetric torsion case will lie within the cross section used to create the body of revolution. Also, it must properly relate to the axis of symmetry chosen by the program. Also, in the axisymmetric case, lines of geometric symmetry become model lines of symmetry.   An axisymmetric shaft of cross section shown in Figure 8-3 may use symmetry to reduce the model to the shaded area. If no lines of symmetry exist to reduce the model, then at least one point must have a fixed value of zero for the twisting function to prevent rigid body motion.

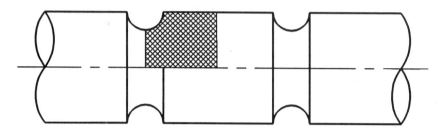

Figure 8-3. An Axisymmetric Shaft in Torsion

We can make a 3-D model of this object by creating a wedge that is cut by two radial planes passing through the axis of symmetry with a small angle between them. We only need one element connecting the planes with mesh subdivision occurring in the cut planes. An example is shown in Figure 8-4. Referring back to Chapter 2, we see that only the circumferential displacement component is nonzero in the solution. Change the nodal displacement coordinate system to the cylindrical system with its origin on the axis of revolution. Set to zero all the radial and axial displacement components of all nodes. Set the circumferential components to zero for all the nodes on one of the symmetry planes. Finally, set the circumferential components to values linearly proportional to radius for all the nodes on the symmetry plane on the other end of the model. This action provides the torsion load. Following solution, we may examine the distributions for the $\tau_{r\theta}$ and $\tau_{z\theta}$ shear stresses to determine the stress concentration factor.

8.3   Computer Input Assistance

Generating the mesh for the prismatic bar torsion has the same requirements outlined for general 2-D plane models or for 3-D models. The mesh generation for axisymmetric torsion has the same requirements outlined in the previous chapter on axisymmetric body analysis. Mesh generation for 3-D analysis of axisymmetric torsion requires solid wedge elements at the core or a very small radius hole with total restraint of nodes near the axis of symmetry when dealing with solid members.

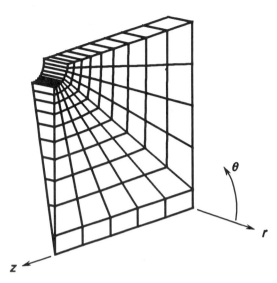

Figure 8-4. A Wedge Section Grooved Shaft 3-D Torsion Model

8.4   The Analysis Step

It is very common that a prismatic bar in torsion also has other load components such as axial load or flexural load. Before resorting to a three-dimensional analysis remember that we can get the axial and flexural stresses from direct P/A and Mc/I formulas. Since the analyses are all linear, we can use superposition to obtain the total state of stress. In case the member is simply a component of a more complex structure, a torsional analysis can determine the torsional stiffness of the member. That in turn is input for a 3-D beam element model that will determine what other stress components exist. The resulting torque load found from the beam element model scales results from the torsional analysis to set the magnitudes of the torsional shear stresses. Superposition of these with the other results give the combined state of stress in the member.

Circular shafts with changing diameter also may have torsional loading in combination with axial and lateral loads. Axial loading is a normal axisymmetric load condition. Analysis of lateral bending loads uses a non-axisymmetric loading capability (if it exists) as does the torsion antisymmetric load. If the torsion analysis is done with a separate element rather than with the axisymmetric element with antisymmetric loading, then the stresses resulting from all the other load conditions superimpose to produce the total state of stress.

8.5   Output Processing and Evaluation

Once again most of the 2-D guidelines apply to these cases, however for the purely torsional cases only the shear stress components will exist. The primary solution variable is either the warping function or the twisting function derived in the displacement formulation. Evaluation of contour plots of these functions compared with the finite element mesh densities should be done to estimate the accuracy of the results. For example, if we use ten contour lines in plots, then for reasonable accuracy there should be no more than one contour passing through each element.

If the postprocessor can produce the three-dimensional displacement field that results from the warping or twisting function and display deformed shape plots, then they also can help verify the reasonableness of the solution. The stress components are difficult to interpret intuitively, but they should definitely match the required boundary conditions.

8.6   Case Studies

The first case will examine the torsion of a prismatic bimaterial bar with a square cross section. The cross section sketch is in Figure 8-5. These results correspond to those in reference [8.1]. The overall cross section measures 10 units by 10 units. Material 1 measures 6 by 10 and the outside strips of material 2 measure 2 by 10. We study three combinations of material property ratio.

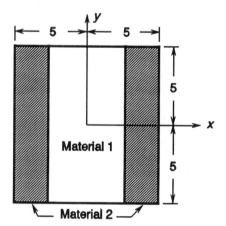

Figure 8-5. Torsion of a Square Bimaterial Shaft

Symmetry allows creation of a quarter section model as shown in Figure 8-6. The boundary conditions are that the warping function be zero on the horizontal and vertical planes of symmetry. This is equivalent to zero values of axial displacement along these edges of the model.

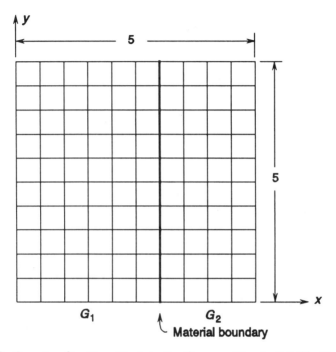

Figure 8-6. Quarter Section Model of a Square Bimaterial Shaft

The contour plots for τ_{xz} are in Figure 8-7. For any material element these are shear stresses acting on the z face in the x direction and on the x face in the z direction. The right edge at $x = 5$ is a free edge and τ_{xz} must equal zero, which agrees with the plots.

First, examine Case 1 where the two materials have the same shear modulus of elasticity. The values increase toward the top center of the cross section to a value of -4.81. The normalization factor on the values is $10^{-3} T/L^3$, where T is the applied torque, and L is the length unit. In Case 2, the inside material 1 is three times stiffer than material 2 and the contours move toward the inner stiffer material. The maximum value of stress is also higher reaching a value of -6.12 at the top center. Case 3 has the outside material 2 three times stiffer than the inside. In this case, the contours move outward with a lower maximum value nearing -4.0. This looks like an improvement by reducing the stress level for the same applied torque. However, upon examination of the τ_{yz} component in Figure 8-8, the maximum here has also increased to 6.43 in Case 3.

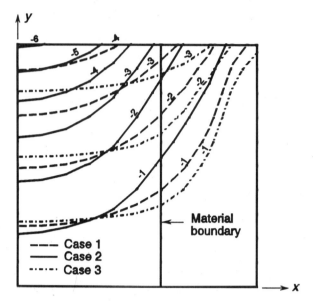

Figure 8-7. Contour Plots of τ_{xz}

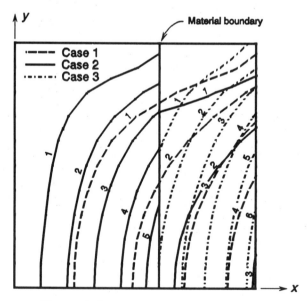

Figure 8-8. Contour Plots of τ_{yz}

The top edge in this plot is a free edge and therefore τ_{yz} must equal zero. Notice that the contours of τ_{xz} and τ_{yz} for Case 1 with homogeneous material are reflective symmetric about the 45° line as they should be. In Case 2 with stiffer material on the inside the contours are drawn inward and the maximum τ_{yz} shows to be near the material interface. Actually, it must be right at the interface. If the interface has a perfect bond, then compatibility conditions require that the γ_{yz} shear strain must be the same on both sides of the interface, and the τ_{yz} stress components must make a step change in proportion to the shear modulus ratio.

The magnitudes must differ by the ratio of the material shear moduli of elasticity. The plots in [8.1] are a result of node stress averaging which smooths out the contour lines and gives a false impression of the stress distribution. The stress distribution of the τ_{xz} shear stress must be continuous across the interface to satisfy equilibrium conditions, but the τ_{yz} component must be discontinuous if the shear modulus changes value.

Another case of a noncircular prismatic bar is a splined shaft with four semicircular keyways sketched in Figure 8-9. These results were also after reference [8.1].

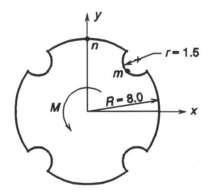

Figure 8-9. Splined Shaft With Four Circular Keyways

An eighth section model is sufficient for the symmetry present, but the authors chose to use a quarter section model as presented in Figure 8-10. The contour plot displays the warping function for this cross section. The values correspond to the magnitude of z direction displacement for input of a unit value of angle of twist of the shaft. The 45° line of symmetry is clearly shown. The maximum shear stress developed at point m was 2.15 times the nominal value at point n.

The final case is one of torsion of a varying circular cross section, a groove in a circular shaft. A drawing of the geometry is given in Figure 8-11 where the shaft has a radius, b, and the groove radius, R, is one-tenth of the shaft radius. These results are after reference [8.2].

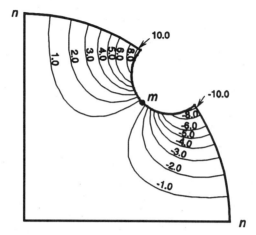

Figure 8-10. Warping Function on a Quarter Section of the Shaft

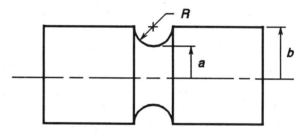

Figure 8-11. Circular Shaft With Semi-Circular Groove

The axisymmetric model is shown in Figure 8-12 where the axis of symmetry lies along the bottom edge, and the half section model uses the symmetry plane on the centerline of the groove. A torsional shear load must apply on the right edge that varies linearly with radius. This corresponds to the distribution that develops naturally in a constant circular cross section shaft.

Results from the analysis are in Figure 8-13. The maximum shear stress is shown along the surface from the bottom of the groove to the end of the model. We normalize the plotted stress by the nominal stress, τ_0, that would exist in a shaft of radius 0.9b at the bottom of the groove. This shows a stress concentration factor of about 1.61 and a stress of $0.73\tau_0$ along the constant cross section. The shear stress at the edge of the groove must be zero because of the orthogonal free surface.

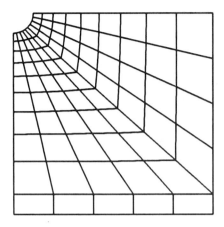

Figure 8-12. Axisymmetric Half Section Model of the Grooved Shaft

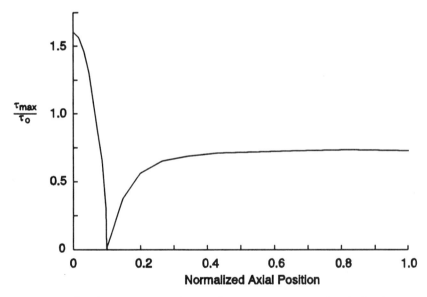

Figure 8-13. Maximum Shear Stress Along the Shaft Surface

Problems

8.1 Various cross section shapes for a member under torsion loading are given in Figure P8-1. For each of these shapes, determine the torsional constant and the location and magnitude of maximum shear stress proportional to applied torque. If the computer program in use does not have a torsion element, then do a 3-D solid analysis with prescribed displacements as discussed in the text. Also, do not apply any axial displacement restraints that prevent the cross section warping except on symmetry planes.   Exact and approximate solutions for some of these shapes are available in references [8.3] and [8.4].

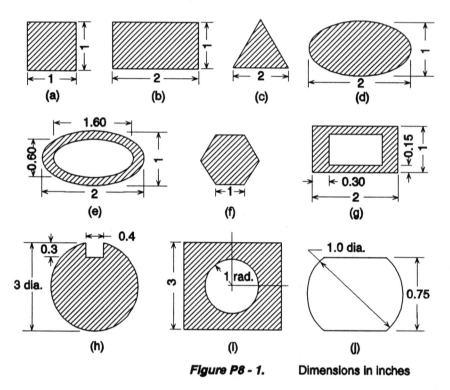

Figure P8 - 1.   Dimensions in inches

8.2   In Figure P8-2 an I section has a plate welded to the right side to increase the torsional rigidity. How much did the torsional constant increase by addition of the plate? Has the maximum shear stress increased with addition of the plate for a given applied torque?

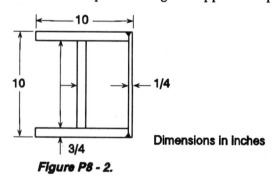

Figure P8 - 2.

8.3 A structural member shown in Figure P8-3 subjected to torsion load is made by welding a structural steel channel section to a steel tube. Determine the torsional constant and location and size of the maximum shear stress in this configuration for a given torque load.

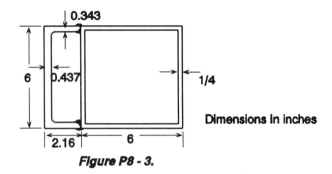

Figure P8 - 3.

8.4 A stepped shaft section with a fillet radius is shown in Figure P8-4. Torsion loading of the shaft produces a stress concentration on the nomimal shear stress in the round shaft. Determine the shear stress concentration factor by finite element analysis and compare with published data for this geometry. If the option to apply circumferential direction loads to axisymmetric solids is not available in the computer program you are using then use a 3-D model with prescribed displacement loading.

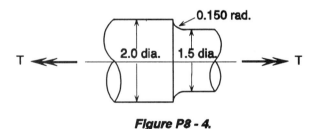

Figure P8 - 4.

8.5 The rotating shaft section in Figure P8-5 has a grinding relief groove of 5 mm radius that is 2.5 mm deep. Determine the stress concentration in the groove due to torsion loading. Compare with any published values available. This is the same geometry as in Problem 6.5, so compare with the 3-D model if it was done and if a circumferential loaded axisymmetric analysis was done here.

8.6   The alternative relief groove geometry to that in Problem 8.5 is shown in Figure P8-6. Determine the torsional stress concentration factor for this case and compare with published data or Problem 8.5 or Problem 6.6.

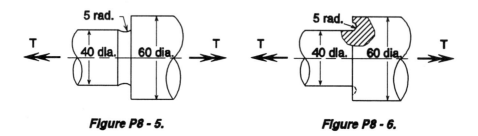

Figure P8 - 5.   **Figure P8 - 6.**

References

8.1   Boresi, A. P. and Sidebottom, O. M., *Advanced Mechanics of Materials*, 4th Edition, John Wiley & Sons, Inc., 1985.

8.2   Fenner, D. N., *Engineering Stress Analysis, A Finite Element Approach With Fortran 77 Software*, Halsted Press, John Wiley and Sons, Inc. 1987.

8.3   Baumeister, T. (ed.), *Mark's Mechanical Engineers' Handbook*, McGraw-Hill, New York

8.4   Cook, R. D. and Young, W. C., *Advanced Mechanics of Materials*, Macmillan Publishing Co., New York, 1985.

CHAPTER 9

THIN-WALLED STRUCTURES

This chapter considers finite element modeling of plate and shell structures. The mechanics formulations for these two classes of structures are significantly different. The membrane effect present in shell structures is absent in linear plate theory. However, many of the current finite element formulations for shells will perform for plate as well as shell structures. We will consider plate and shell structures together as a class of thin-walled structures.

This area of finite element analysis has historically been the most difficult to achieve reliable and cost effective solutions. Further, there is not now any particular plate or shell element formulation that is broadly acceptable to the analysis community. There has been an enormous amount of research devoted to plate and shell element formulations. One recent publication [9.1] listed a catalog of 88 different plate element formulations. The *Finite Element Handbook* catalogs and reviews nine classes of plate elements [9.2] and nine classes of general shell elements [9.3]. There are probably not quite as many shell elements as plate elements because they are more difficult to formulate. Obviously the reason for the existence of many different elements is that we have yet to find an element which gives completely satisfactory results. So the question is, what is an engineer to do?

The first answer must be to proceed cautiously and never trust the solution results without considerable effort to verify the correctness of the model as well as the computer algorithms and solutions. Further you should corroborate the results with your experience, any experimental data collected on the project, and other closely related solutions.

You also might ask, why not develop fully 3-D solid models of the analysis problem? The answer to this question is the limitations of the computing capability and consequent time required to perform such an analysis. The model would require a five-to-ten element subdivision through the thickness, and so all the element dimensions throughout the model would be on the order of one-fifth to one-tenth of the wall thickness of the structure in order to preserve reasonable aspect ratios and prevent serious numerical ill-conditioning. For thin-walled structures, the number of nodes and elements required, in even a simple model, would be prohibitive.

Since computer capacity is growing rapidly, there may come a time in the future when a three-dimensional model might be the better approach, but at this point practicality demands that we use plate and shell elements. Plate and shell theories are complex, but they are still a simplification of 3-D elasticity solutions. This chapter will not delve very deeply into any specific element formulation, but try to point out the common features of element formulations and the knowledge required to use them in a model.

9.1   Plate Element Formulations

The coordinate system symbol conventions for finite elements are usually different from the conventions used in classical plate theory. So before discussing element formulations, we should define the symbol conventions. A quadrilateral plate element shown in Figure 9-1 has a local coordinate system with the plate element lying in the x-y plane. On one node, the node force components are shown and on another node the node displacement components are shown. In Chapter 2 the displacement components for a plate element were the lateral or w translation and its derivatives with respect to the local x and y directions. For small displacements these derivatives are equal to the rotation angle of a normal to the plate in the x and y directions.

Label the rotation vector, $(\partial w / \partial y)$, pointing in the x direction, θ_x and the rotation vector pointing in the y direction θ_y. The force components that correspond to these displacements are the lateral force, f_z and the moments, m_x and m_y. The conventional plate element then will have three degrees-of-freedom per node.

In classical plate theory these rotation and moment components typically carry the opposite subscript notations [9.4]. Therefore when comparing plate solutions from textbooks or handbooks to finite element solutions be sure to check the component symbol definitions.

Plate element formulations may broadly classify into Kirchhoff, Discrete Kirchhoff, and Mindlin elements [9.5]. These classifications associate the element with Kirchhoff thin plate theory and Mindlin thick plate theory. Discrete Kirchhoff elements attempt to improve the performance of Kirchhoff elements by satisfying displacement and slope compatibility only at discrete points on the element.

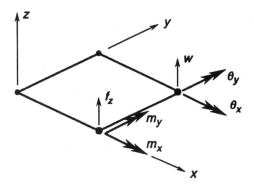

Figure 9-1. Plate Finite Element

The earliest elements formulated derive from Kirchhoff classical plate theory. This theory requires C^1 continuity of the displacement field, meaning that the lateral displacement, and its first derivative with respect to the two in-plane coordinate directions, must be continuous across element boundaries to satisfy compatibility conditions. Beginning with corner noded triangular elements, the researchers found it impossible to derive a polynomial displacement interpolation that could guarantee C^1 continuity using only the three geometric degrees-of-freedom (DOF). The three DOF being the lateral displacement and its two derivatives at each node. Researchers have tried many interpolation schemes for triangular and quadrilateral Kirchhoff elements to provide the reliable formulation needed for this type of plate element, but none has yet been completely satisfactory [9.6]. Mostly, these corner noded low-order Kirchhoff elements are no longer of interest.

Elements that have been successful for classical Kirchhoff theory use higher order interpolation schemes. These elements use midside nodes and 18 to 21 total degrees-of-freedom involving lateral displacement and first and second derivatives at selected node points. They have been relatively successful at producing good solutions, but due to the many DOF for each element the element computations are costly and the resulting system of equations has a large bandwidth.

More of the recently formulated plate elements use the Mindlin (thick) plate theory for a basis which includes the shear deformation that becomes significant in thick plates [9.6]. The primary difference between Kirchhoff and Mindlin plate theories is that Mindlin theory does not require normals to the plate to remain normal after deformation. The rotation of the normal is then equal to the slope plus the rotation due to shear deformation as shown in Figure 9-2.

We can not quantify the difference between a thin and thick plate, but can only state that the deformation produced by the transverse shearing stresses is significant in thick plates. From a finite element displacement

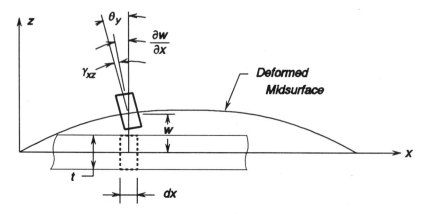

Figure 9-2. Mindlin Plate Normal Rotation

formulation point of view this allows the lateral displacement and rotations to be independent and therefore the element formulation only requires C^0 continuity or compatibility of the lateral displacement and rotations along the element boundaries. This greatly simplifies the displacement interpolation schemes employed to formulate the plate element. However, the approach is not without pitfalls either [9.6].

The displacement formulations based on Mindlin plate theory must interpolate for the following displacement components.

$$w, \ \theta_x, \text{ and } \theta_y \tag{9.1}$$

These displacement components may be rather easily interpolated and a four node isoparametric quadrilateral element formulated [9.5] through

$$w = [N]\{w_i\}$$
$$\theta_x = [N]\{\theta_{xi}\} \tag{9.2}$$
$$\theta_y = [N]\{\theta_{yi}\}$$

with geometric interpolation in the local x-y element plane given by

$$x = [N]\{x_i\}$$
$$y = [N]\{y_i\} \tag{9.3}$$

The formulation of the element continues by writing the strain-displacement relations and stress-strain relations and following the outline presented in the chapter on three-dimensional solids. However, the strain component, ϵ_z, and stress component, σ_z, which act through the plate thickness are small enough to neglect.

A problem that occurs in this formulation happens when the element aspect ratio of in-plane length to thickness becomes large. While the element represents a relatively thick plate, that is, a low aspect ratio, the transverse shear strains are small but significant compared with the other strains and the element behaves well. However, as the aspect ratio becomes larger and the element becomes more of a thin plate, then the strain energy due to the transverse shear stress and strain becomes artificially large and may result in mesh "locking"[9.6]. This locking causes the displacements and stresses in the model to become ridiculously low for the given applied load.

In this element, integration for the stiffness is done numerically by Gauss quadrature, and in the case of a corner-noded element, two-point Gauss integration is sufficiently accurate. One remedy for mesh locking is to use a reduced order of integration for the portions of element stiffness due to transverse shear. If using two-point quadrature for the bending stiffness components, then use one-point quadrature for the transverse shear stiffness component. This allows the element to behave much better as the aspect ratio goes higher, but it will still eventually lock at higher aspect ratios. Another remedy is to reduce the element size in the model.

Despite this problem the isoparametric Mindlin theory element is usually chosen over elements formulated to guarantee C^1 continuity by the Kirchhoff theory. Other approaches to plate element formulation have used different interpolation formula to compute the transverse shear strains from the nodal values of displacement and rotation and many other schemes to avoid the locking problem. There are also several approaches for higher order elements using midside nodes and selective reduced integration.

As long as the problem we are trying to solve fits the assumptions of classical plate theory we can achieve a solution with an element described as above which only has one lateral displacement and two rotational degrees-of-freedom. This kind of problem or structure is one in which the member lies in a plane, has loads normal to the plane, and has supports against lateral motion. However, when our real structure consists of flat sheets of material formed into box shapes or other 3-D forms as shown in Figure 9-3, then the inplane or membrane effects may also be important.

An additional effect occurs whenever a loaded plate deforms into a more shell-like structure with curvatures and therefore develops in-plane strains and forces. These in-plane forces carry part of the load, and as the load increases the membrane forces increase more rapidly and the plate becomes stiffer. This is a geometrically nonlinear problem, and in order to solve that problem, we must include the membrane effects within the element formulation. In some cases the linear classical plate solution may

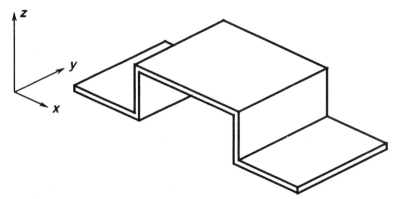

Figure 9-3. A Three-Dimensional Plate Structure

be in error by as much as 50 percent when the lateral displacement is equal to the thickness.   So to solve nonplanar plate problems and nonlinear problems the plate element formulation must include membrane components.

In plate elements, inclusion of the membrane components may simply amount to superposition of the 2-D plane stress element onto the plate element. This results in five DOF per node as shown in Figure 9-4. The rotation DOF about the local z axis, often called the "drilling" DOF [9.6], is absent in the formulation. However, to couple plate elements to make a nonplanar model this rotation DOF must be present. The typical approach is to assign a small fictitious rotational stiffness to this DOF. This makes it active so it will couple to out-of-plane elements and will not cause a singularity in the structure stiffness matrix. However, we cannot expect a correct response if we connect a beam element normal to the plane to a node and load it with a torque since the stiffness value is arbitrary.

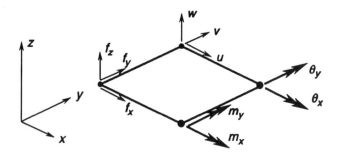

Figure 9-4. A Membrane and Bending Plate Element

9.2   Shell Element Formulations

Shell structures always develop some membrane forces because of the initial curvature, and therefore any shell element formulation must include the membrane components.

The formulation of shell elements may take at least three different routes [9.5]. One, combine a flat plate element with a plane stress membrane element by simple superposition to provide a flat shell element. Two, the element can be a curved element formulated on the basis of a shell theory. Three, a 3-D solid parabolic isoparametric element may be degenerated to match the behavior of the Mindlin plate element.

The first route is conceptually simple and easy to formulate, but has not proven to be satisfactory in many cases. The second route usually results in an element with 30 to 60 DOF involving normal displacements, and its first and second derivatives and usually cubic interpolation of in-plane displacement components. Some of these elements have shown to be accurate, but they are very expensive and time consuming to employ on practical models. The third route is in favor among some active researchers, but as yet the development is not complete.

The flat element with superimposed membrane element has five DOF, the three linear displacement components and two rotation components lying in the plane of the element in each element's local coordinate system. When these elements connect to model a curved surface illustrated in Figure 9-5, the stiffnesses with respect to the local element coordinate system transform to global components for summation. The arbitrary small value of the torsional or "drilling" stiffness normal to the element is not a factor since the direction of this normal changes from one element to the next in a curved shell model.

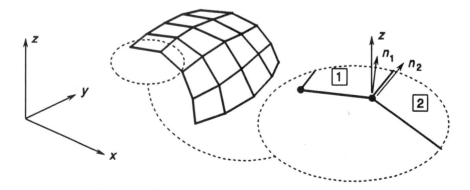

Figure 9-5. A Curved Shell Modeled With Flat Elements

The isoparametric shell element formulation proceeds in a manner similar to the plate element formulation, however the displacement components will consist of the three linear displacement components for all points on the midsurface of the shell and the two rotations utilized in the plate formulation. The stress and strain components, σ_z and ϵ_z, are dropped again so that we are left with five components of each. These include the transverse shear strains and stresses and all the membrane and bending normal components. This formulation behaves well but has the same problem with aspect ratio that the plate element has in that as the shell thickness becomes smaller the aspect ratio goes higher and the element tends to lock. In the shell element this is caused from the increase in strain energy of the shear components and the membrane components as well. The typical fix for this problem is again to underintegrate or use a lower order integration for the membrane stiffness and transverse shear portion of the element stiffness matrix. This allows the element to perform better with thinner elements.

Again in this element there is no stiffness associated with the rotation or drilling DOF normal to the midsurface. Therefore, if we model a flat section of the shell with multiple elements lying in the same plane, then a small artificial stiffness avoids a singularity in the structure stiffness matrix.

9.3   The Finite Element Model

If the chosen computer program has several element types available to model thin walled structures, then the user must choose which element to use for the type of problem to be solved. Because of the many difficulties involved in plate and shell element formulation the program documentation may, or more likely, may not fully describe the formulation and features of each element available for use. Therefore, the user may not be able to make a good judgment about which element to use.

Typically, plate and shell elements do not completely satisfy compatibility conditions along element edge connections. An element that does satisfy compatibility will converge to the exact solution as we refine the mesh. An element that does not satisfy compatibility may or may not converge to the correct solution. Something to look for in an element description is if the element will pass a "patch" test [9.7]. If it does pass the patch test then it will converge to the correct solution as we refine the mesh.

If it is not known whether it passes or not, then the user should perform patch tests on any element before using the element in any specific analyses. You can find the details of performing a patch test in more advanced finite element analysis textbooks. Basically, the purpose of the test is to see if the element will correctly respond to applied uniform states of stress and allow rigid body motion without creating any stress. The patch test may check any type of element, however, with plate and shell elements in particular, some elements may pass the patch test in some mesh layouts and may fail in others.   So, we should perform multiple tests on any element under consideration for use [9.8, 9.9]. Passing the patch test does guarantee convergence to the correct solution as we reduce the mesh size as long as numerical round-off errors are minimal, but it does not guarantee any particular rate of convergence.

If the structure under analysis fits the classical plate definition in that it lies in a single plane with only lateral loads, then a simple plate element is a proper choice.   In this case only the lateral displacement and the two rotation components in the plane of the midsurface are necessary DOF in the model. If the choice is to use triangular plate elements, then the mesh arrangement in Figure 9-6(a) performs better than arrangements in Figure 9-6(b) or Figure 9-6(c).   A corner noded element, whether triangle or quadrilateral, is the right choice for the initial models.   When satisfactory convergence with the linear elements is apparent then perhaps a higher-order element model will confirm the results.   If Kirchhoff (or thin plate type) elements are chosen, then the higher-order elements usually will produce more satisfactory and confident results.

(a)   (b)   (c)

Figure 9-6. Triangle Plate Element Mesh Arrangements

If the structure geometry consists of flat panels connected in more than one plane as shown in Figure 9-7, then the simple 3 DOF plate element may not accurately represent the structure's response.   Specifically the user must realize that the results at the connections of nonplanar panels will not be an accurate representation, since the membrane or in-plane displacement components are absent from the formulation and therefore offer no resistance against any displacement in the plane of any of the panels.   In these types of structures it is better to use a plate element which has the membrane formulation included or a shell element.   Of course, any thin-walled structure that has curved surfaces as shown in Figure 9-8 must use shell elements in the model.

In many programs the shell element is always the better choice because any plate elements available may be of old formulations which do not

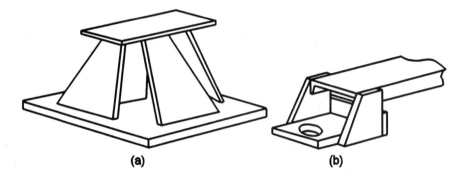

Figure 9-7. A Three-Dimensional Plate Structure

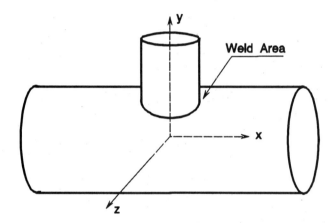

Figure 9-8. A Three-Dimensional Shell Structure

perform well. Since research is now focusing more on shell elements the shell element may be a more current formulation. If the shell element is of a more recent formulation then it may substitute for a plate element with 5 DOF per node. Also, it could be a simple 3 DOF plate element by constraining the 2 DOF associated with in-plane displacements of the midsurface.

As always, the mesh plan should provide adequate subdivision for any expected variations in the stress field. However, in areas where panels join or the surface makes sharp radius turns, the structure may not behave according to plate and shell theory assumptions. For the examples shown in Figure 9-9, we may find it necessary to analyze the structure with an overall model of plate or shell elements, then follow that with a local analysis using 2-D or 3-D solid elements. For this case, plan the mesh so that a local group of plate or shell elements will provide a suitable set of boundary information for the local solid element model. This approach is discussed in Chapter 11.

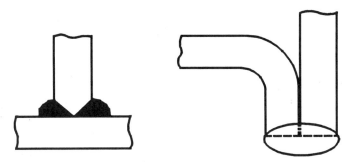

Figure 9-9. Panel Connections and Shell Sharp Radius Turns

Many plate and shell structures are made by an approach in which the plate or shell member is like a skin with reinforcing beams attached to one side of the skin to provide greater rigidity to the overall structure. The mesh plan should provide lines of nodes and element edges along paths of reinforcing beam attachment. The reinforcing beams then become part of the model by the addition of beam elements along the line of nodes to represent the reinforcing beams. Since the beam centroid usually does not coincide with the plate or shell midsurface, there must be some capability to offset the beam element centroid from the model nodes lying on the plate or shell midsurface. This is shown in Figure 9-10. There may be a significant error associated with this coupling of beam and plate elements especially for coarse meshes [9.10].

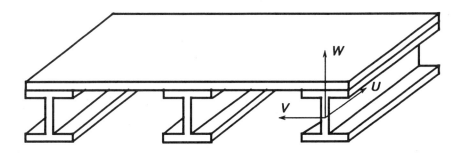

Figure 9-10. Reinforced Plates and Shells

We should continue to exploit symmetry conditions as much as possible to keep the model size under control. In the case of plate and shell structures, symmetry enforcement depends on the restraint of both translational and rotational DOF. For example, a square plate with uniform pressure loading and simply supported edges is shown in Figure 9-11. Symmetry of the geometry and loads suggests that a quarter section of the structure is adequate for the model. Then to enforce symmetry, restrain the nodes along the symmetry lines to prevent any in-plane displacement of the midsurface perpendicular to the symmetry line as well as any rotation that would cause a material particle above or below the midsurface to move off the symmetry plane.

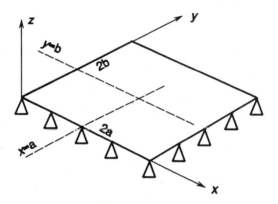

Figure 9-11. A Square Plate With Uniform Pressure Load

In this case restrain nodes along the line x = a against u displacement and θ_y rotation. Restrain the nodes along line y = b against v displacement and θ_x rotation.

Displacement restraints also provide support boundary conditions. Restraint of vertical displacement, w, provides for the simply supported edge shown in Figure 9-11. Also it would be logical to restrain the rotation component normal to the edge. However, this tends to overrestrain the edge in plate and shell elements and is not necessary since the vertical displacement between nodes on corner noded elements interpolates linearly.

This problem is even more severe in plate structures with curved edges modeled with straight edged elements. Look at the simply supported curved edge in Figure 9-12. If we restrain the rotation normal to the edge on two adjacent elements with noncolinear edges, then that will automatically lock both rotation components at the common node. In fact, the edge is now clamped rather than being simply supported [9.6]. This is a serious problem in some models that we avoid by allowing both rotation components to remain free and only restraining vertical displacement. This will converge to the true simply supported boundary condition as we refine the mesh along the edge.

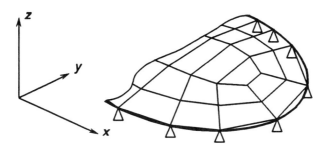

Figure 9-12. A Simply Supported Curved Edge Plate

Loads apply in the conventional manner as described in previous chapters in that they must apply at nodes or the program must convert pressure or body forces to nodal loads. However in plate and shell elements, loads may be either forces or moments or both. Typically, distributed loads such as pressure or acceleration loads will result in a statically equivalent set of both forces and moments at the element nodes.

For a 3 DOF per node simple plate element, remember that only forces normal to the plane and moments parallel to the in-plane rotation vectors may apply. Shell elements or more general plate elements also may carry inplane forces.

9.4   Computer Input Assistance

Mesh generation for plates and shells can normally be done with a 2-D area mesh generator available in most programs. Obviously, for conventional plate structures the 2-D area mesh generation will work. A mesh for a plate element model will look the same as a 2-D plane model or a 2-D axisymmetric model. The selection of a plate element instead of a 2-D element type makes it a plate structure model. For three-dimensional plate structures we may use the area mesh generator in each panel of the structure to generate an area mesh that we can connect as in the parent structure. An example of a three-dimensional plate structure mesh is drawn in Figure 9-13.

For curved shells many program preprocessors allow forming of an area mesh onto the defined surface of the shell. If the shell has a closed volume or is relatively deep with respect to its surface area, then it may be necessary to define subareas of the shell that are easier to form with area meshes. This helps avoid severe distortion of elements caused by excessive stretching or draping of the mesh. An example of a deep shell mesh is drawn in Figure 9-14.

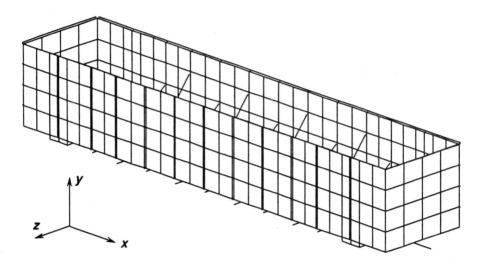

Figure 9-13. A Three-Dimensional Plate Mesh

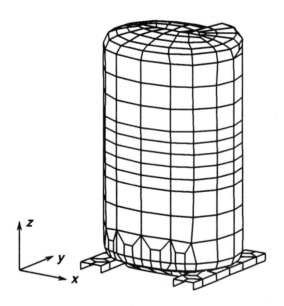

Figure 9-14. A Deep Shell Mesh

Model checking is done much the same as in other finite element models to check for continuity of the mesh plan, proper geometrical representation, and minimal element distortion. Because of the difficulties in formulating a good plate or shell element, most of the formulations in current use are highly sensitive to distortion. The user should make every effort to develop a regular mesh of near square elements whenever possible. Automatic application of boundary conditions and loads are typically done in the same manner as for other finite element models. However, the graphic display of boundary conditions and loads on plate and shell element models may become very visually cluttered. Typically each node may have up to six DOF, three translations and three rotations, and display of these vectors at each node that has displacement restraints and loads becomes difficult to see.

We must observe the graphic display from several viewpoints in the case of three-dimensional plate or shell element models to visually check the entire model. The mesh will have the appearance of a "fishing net" for curved shells or a "lattice framework" for a three-dimensional plate structure since the material thickness is an element physical property that does not show in the geometric display.

In these models it is critical to employ bandwidth and wavefront minimization because of the usual large area of connection of elements and the fact that most of these elements have many DOF per element. So to gain the best possible numerical performance it is imperative that the bandwidth be minimal.

9.5   The Analysis Step

The complexities of plate and shell element models for any significant size problem generate a large set of equations. They will, because of the element formulations, have a rather large bandwidth and thereby become difficult to solve accurately in a reasonable time period. The equations also can become ill-conditioned rather quickly especially in elements that include in-plane membrane effects with the bending components. This is because the membrane stiffnesses are much larger than bending stiffnesses. These problems make it difficult for the user who wants to develop analyses for real problems to achieve satisfactory results.

Most of the research effort focuses on element formulations and convergence criteria, but the test cases for evaluation are usually small sized academic cases rather than actual problems. So the demonstrations of adequacy in the research efforts may not be very helpful in the practical cases where an analysis must be done. All of this points out that the user must be extremely wary and exercise careful judgment about the model and the results obtained from the use of that model.

The solution step may fail due to several factors other than the normal model errors such as highly distorted elements or allowing rigid body motion to occur. If the equations are too ill-conditioned, then the solution will be completely invalid and usually recognizable as trash. If this occurs, then it may help to refine the mesh. The bending stiffnesses are proportional to the cube of the element's average side length while the membrane stiffness is linearly proportional to the side length. Refining the mesh then will bring the stiffness values closer together even though there will be more total system equations. If the solution trashed because of round-off error, then refinement will make it worse. In this event, we must simplify the mesh with fewer total equations.

If using the Mindlin thick plate element or its shell element counterpart and mesh locking occurs, it should be easy to recognize. The displacements and stresses reported will be extremely lower than expected results and probably will bear no resemblance to the true solution. Refining the mesh also usually solves this problem. Making the element length smaller with respect to its thickness will reduce the strain energy imbalance caused by fictitious shear strain energy in elements with a high aspect ratio. The locking threshold aspect ratio for these elements usually passes quickly rather than by a slow increase in error as the aspect ratio increases. So it may not take much mesh refinement to unlock the mesh.

9.6   Output Processing and Evaluation

The complexity of these problems and the large amount of data output makes it impractical to derive much from printout listings. Evaluate the graphic display of the structure responses by plotting the deformed shape and checking for the proper application of boundary conditions and the expected pattern of deformation of the structure. Again, for three-dimensional plate or shell structures many different views are necessary to make a proper evaluation. It is very helpful if hidden-line plots are available to avoid the visual clutter.

Scan the magnitudes of displacement over the model to see if any are nearly equal to or greater than the material wall thickness. If they are, then there is a strong probability that the solution has some significant nonlinearity. In this event the linear solution error may be very high. You should check for solutions in the literature that have some similarity to the current problem and determine if the nonlinear response is strong for the literature case. If it is, then a nonlinear solution of the current problem is necessary. If not, then depending on how close the similar solutions are to the current problem, a nonlinear analysis may not be necessary. If the accuracy of the analysis is critical, then you should run the nonlinear solution when the linear displacements are nearly equal to or greater than the thickness.

The stress component plots in most post processors are available on the bottom, middle, or top surface of the element. Selection of the midsurface display will plot only the membrane distributions. These can be useful when evaluating the structure for limiting failure loads allowing local yielding at the surface. Selection of either the top or bottom surface produces stress plots of the combined bending and membrane stress on the two surfaces. These locate the maximum stresses which occur in the plate or shell structure.

Postprocessors also allow selection of coordinate systems in which to plot stress components. In conventional plate structures that lie in one of the global planes, the global stress components such as σ_x or σ_y will be the components lying in the plane of the plate. However, in three-dimensional plate models or curved shell models the global components will have very little meaning since they are not in the plane tangent to the surface. In these cases, it is necessary to request the components in local element coordinates tangent to the surface.

As mentioned in the section on element formulation, some of these elements may have difficulty providing convergence to the correct solution and some may not converge at all and others to the wrong solution. The elements that pass multiple patch tests in various configurations will converge to the correct solution, however the convergence may be slow and it may be oscillatory rather than monotonic. In these structures, since there are very few analytical solutions available, it is also very difficult to have good approximate solutions for backup. To have high confidence in the validity of the solution, convergence becomes extremely important to assure accuracy. Therefore, although it becomes a rather tedious job, the analysis iterations should continue until the analyst becomes confident of convergence, otherwise the results reported may be in serious error.

Mesh refinement follows the same basic guidelines presented earlier for other element types. However, mesh refinement must be done very carefully with these models because the chance of large bandwidth, ill-conditioning and other factors increases quite rapidly with the additional DOF added to the model for element subdivision. It may become necessary to settle for a converged solution over the large portion of the model, and then resort to subregion model analysis of local areas to reach a solution. The subregion modeling concept will be discussed in a later chapter.

Failure criteria, as applied to plate and shell structures from the linear static analysis, will involve study of the Von Mises equivalent stress contours on the top or bottom surfaces of the plate or shell structure. However, this is sound only as long as the linear elastic solution is good. If the deformations are too large for the linear solution to be good, then a nonlinear solution must be done to have any assurance of a useful solution. In most plate and shell structures the threshold of large displacement nonlinearity begins when the displacement of a point normal to the surface tangent plane exceeds the material thickness.

Another distinct failure mode for thin structures is elastic instability or buckling under compressive loads. To find the instability loads we must follow a different analysis procedure. Stability analysis is beyond the scope of this text. Instability is very hard to predict, so high factors of safety should be applied to any predicted values.

9.7   Case Studies

The first case is a square plate with simply supported edges and has a uniform lateral pressure load. The plate is steel with a 0.1 in. thickness and has a side length of 10 in. The pressure load is 10 psi. A quarter section model of four elements as presented in Figure 9-15 will be the initial model. This problem has an analytical solution from classical plate theory [9.4]. That solution will be used to examine the convergence of the plate finite element solution.

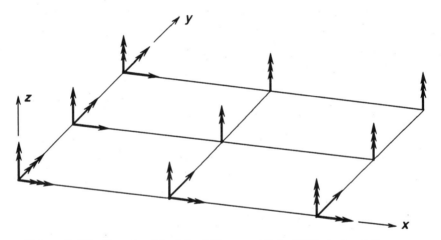

Figure 9-15. Quarter Model of Square Plate With Lateral Pressure

The center of the plate is at the leftmost corner of the mesh. Displacement boundary conditions model the supports and the symmetry restraints. At the nodes the single head arrows signify a translation restraint in the direction of the arrow, a double headed arrow signifies a rotation vector restraint, and a triple headed arrow indicates both the translation and rotation restraint in the direction of the arrow. We restrain the outer edges of the model against vertical displacement to provide the simple support condition. On the edge of symmetry along the x axis the y translation, and the θ_x and θ_z rotations are restrained. On the edge of symmetry along the y axis the x translation, and the θ_y and θ_z rotations are restrained. We may apply or ignore all the θ_z rotation restraints because there is no actual stiffness for that DOF in the element.

The deformed shape of this initial model appears in Figure 9-16. The boundary conditions seem to be proper, and the maximum displacement at the center is 0.132 in. The analytical result is 0.148 in., an error of -7.4 percent.

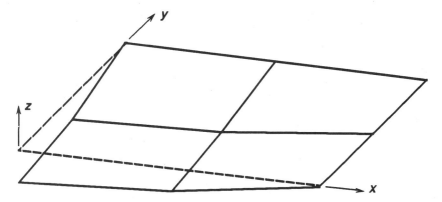

Figure 9-16. Deformed Shape of a Square Plate With Lateral Pressure

The contour plot of σ_x is shown in Figure 9-17. The maximum value of 26.8 kpsi is 8 percent below the analytical value of 28.7 kpsi. The value of σ_x along the right edge must equal zero, and this is nearly correct based on the contours. However, the contour lines change direction too abruptly for the expected stress variations.

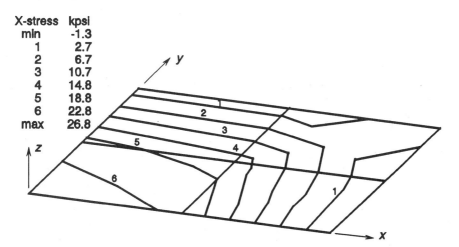

Figure 9-17. Contour Plot of σ_x in a Square Plate

A second mesh comes from doubling the mesh density to 16 elements. The maximum deflection computed was 0.145 in. for an error of -2 percent, and the maximum stress was 28.0 kpsi for an error of -2.6 percent.

The third mesh produced the deformed shape illustrated in Figure 9-18 and has a center deflection of 0.148 in. which agrees exactly with the analytical solution. The stress plotted in Figure 9-19 gives a maximum of 28.4 kpsi for an error of -1 percent. Notice that the contour lines are much smoother than in the four element model.

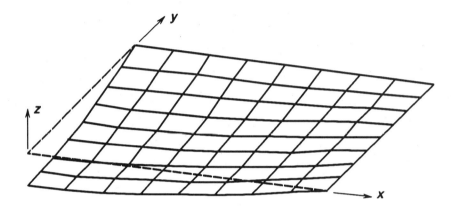

Figure 9-18. Refined Deformed Shape of the Square Plate

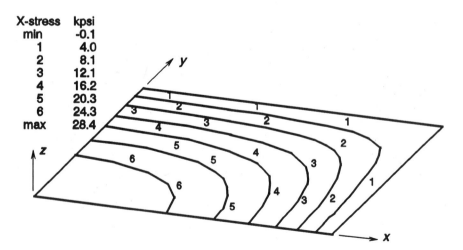

Figure 9-19. Refined Contour Plot of σ_x in a Square Plate

The stress plot is intuitively correct with the maximum occuring at the plate center. However, a check of the Von Mises equivalent stress shows an apparently illogical result. The contour plot in Figure 9-20 shows that the maximum equivalent stress occurs at the outer corner rather than the plate center. This occurs because there is a large shear stress produced in the corner area as the center deflects downward, and creates the highest equivalent stress in this odd location.

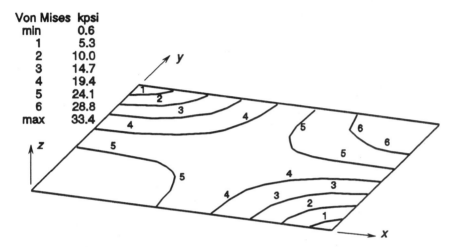

Von Mises kpsi

min	0.6
1	5.3
2	10.0
3	14.7
4	19.4
5	24.1
6	28.8
max	33.4

Figure 9-20. Contour Plot of Von Mises Equivalent Stress

The second case examines the response of a cantilevered beam with a channel cross section to an end load. A standard 6 in. steel channel 36 in. long has a rigid mount at the left end shown in Figure 9-21. This figure displays the finite element mesh generated using shell elements to represent the web and flanges of the channel. The web elements have a thickness of 0.200 in. and the flange elements are 0.3125 in. thick. A load of 2400 lbs. applies at the end across the top flange.

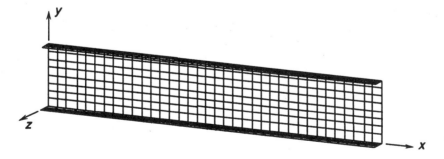

Figure 9-21. Finite Element Model of a Channel Section Beam

Upon solution, the deformed shape of the beam is shown in Figure 9-22. Instead of the expected vertical displacement, the end is deflecting vertically and twisting about the x axis. While the graphic is an exaggeration of the size of the displacements, in this case the vertical deflection at the end is -0.184 in. and the horizontal deflection is 0.376 in. at the top and -0.376 in. at the bottom. There is about twice as much horizontal motion as vertical motion. Furthermore, the conventional beam solution predicts a vertical deflection of only -0.086 in.

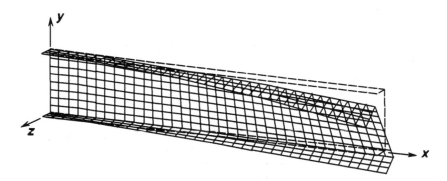

Figure 9-22. Deformation of the Loaded Channel Section Beam

The flexural stress, σ_x, is shown in the contour plot of Figure 9-23, and the zoom view in Figure 9-24 near the mounted end. Overall the contours are reasonable for the kind of distribution expected in a cantilevered beam loading. However, there is a lot of variation across the flange width as shown in the zoom view. The maximum value reported is 54.1 kpsi. A conventional beam calculation predicts 19.0 kpsi which is almost three times lower.

X-stress kpsi
min   -54.1
  1   -38.6
  2   -23.2
  3   -7.7
  4   7.7
  5   23.2
  6   38.6
max   54.1

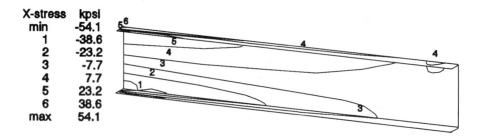

Figure 9-23. Flexural Stress, σ_x, in the Beam

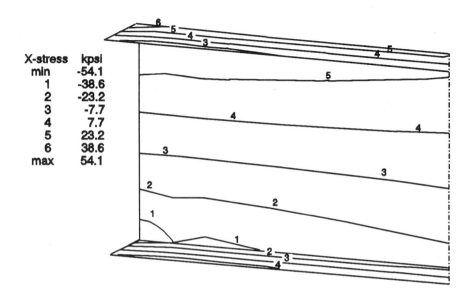

X-stress kpsi
min   -54.1
1   -38.6
2   -23.2
3   -7.7
4   7.7
5   23.2
6   38.6
max   54.1

Figure 9-24. Zoom View of the Flexural Stress

This a disturbing disagreement which suggests the analysis is in error or the beam behavior is really this way and conventional beam theory is inadequate for this problem. It turns out that we have overlooked an important consideration here in that beams with nonsymmetric cross sections must have their load applied through the shear center of the cross section to avoid this torsional reaction of the beam. The shear center concept is usually presented in introductory mechanics of materials textbooks with methods for locating the shear center for several cross section shapes.

There is also an approximate solution for the stress presented in some advanced mechanics of materials texts such as reference [9.4]. It involves several assumptions about the distributions of the shear and normal stress components. The result is a combination of the stress due to flexure with the stress that is generated by the torsion due to the load offset from the shear center. With this solution the predicted maximum value for σ_x is now 27.3 kpsi which is still two times lower than the value from the finite element solution. So we must explore this problem further.

If we add a small plate to the end of the channel beam to apply the load through the shear center, the torsion should disappear, and agreement with beam theory should be better. After doing this, the deformed shape of the beam is shown in Figure 9-25. The twisting is gone and the vertical deflection at the end is 0.0923 in. which is about 7 percent above the beam theory value.

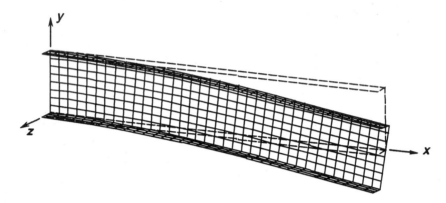

Figure 9-25. Deformation of the Beam Loaded on the Shear Center

The flexural stress, σ_x, contour plot in Figure 9-26 displays contours in better correlation with beam theory. Near the support the stress has some variation across the flanges that beam theory does not predict, but overall the agreement is reasonable. The maximum value is 24.2 kpsi which is about 25 percent greater than the beam theory value. This is not unusual for thin flanged sections.

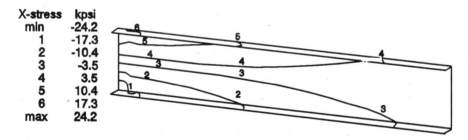

X-stress	kpsi
min	-24.2
1	-17.3
2	-10.4
3	-3.5
4	3.5
5	10.4
6	17.3
max	24.2

Figure 9-26. Flexural Stress in the Beam Loaded on the Shear Center

In this case study conventional beam theory is shown to have some error for flanged beam sections, even when the load acts at the shear center. Also, there is a very large effect when load on the unsymmetrical section does not act at the shear center. This effect is much higher than we predict using approximate mechanics of materials solution methods.

Problems

9.1 The L-shaped plate in Figure P9-1 is 0.100-in. aluminum. Determine the stresses and deflection of the corner caused by the corner load shown. How well can you predict the stresses and deflection with simple beam formulas. Hint: Split the load between two cantilever beams of the same length as the edges of the plate in proportion to their relative stiffness.

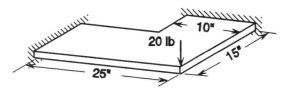

Figure P9 - 1.

9.2 The cantilever loaded plate in Figure P9-2 is 4 mm thick steel. It is approximated in (a) as simply supported along the end and across the plate center with a line load acting on a central 300 mm length of the opposite end with a magnitude of 1.0 N/mm. The actual supports for the plate are two 40 x 40 x 4 mm structural steel angles as shown in (b). Model the plate both ways, (a) and (b), and determine if any significant differences exist in the calculation of stresses and displacement.

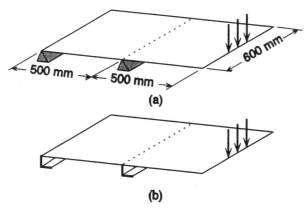

Figure P9 - 2.

9.3   A compressor flap valve shown in Figure P9-3 is 0.018-in. thick hardened steel. The valve operates by deforming into the cylinder space shown on the left as pressure acts on the top annular surface defined by the diameters on the right sketch. The cylinder walls provide vertical simple support under the valve ears at a width roughly equal to the outer diameter of the valve. Determine the stress and deformation patterns of the valve.

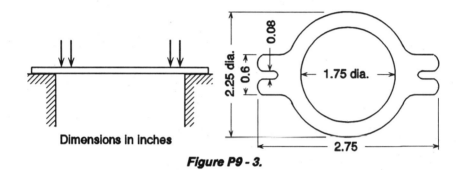

Dimensions in inches

Figure P9 - 3.

9.4   The Belleville spring analyzed by axisymmetric solid elements in Problem 7.7 is shown again in Figure P9-4. It is 0.050-in. steel and has a line load along the upper edge. The lower edge is supported vertically. Model this problem again using axisymmetric shell elements or a quadrant of regular shell elements as illustrated by the coarse mesh in the figure. Determine the initial stiffness of the spring and explain how you would check for significant nonlinearity in the spring for a given appled load range.

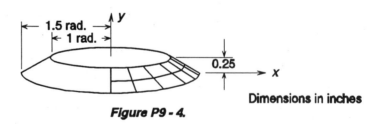

Figure P9 - 4.

9.5   A 3-in. C-channel structural steel member is shown in Figure P9-5. The flanges are 1.5 in. wide and the flanges and web are 0.250 in. thick. Fix one end of the beam and apply a load in the y-direction on the other end and then calculate the location of the shear center for this beam. Hint: The total x-direction moment due to shear flow in the web and flanges at the fixed end will equal the moment of the y-force acting through the shear center, and the free end will not rotate when the y-force acts through the shear center.

Also, determine the torsional stiffness of this beam section when all displacement components on the fixed end are set to zero, and then when the x-displacement components are set free except for one restraint to prevent rigid body motion.

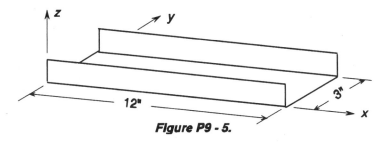

Figure P9 - 5.

9.6   A simple structural steel I-beam is shown in Figure P9-6. Using shell elements, model a 20 ft. long, 12 WF 65 wide flange I-beam that carries a 20 kip vertical central load. Assume that the I is vertical and that the ends are simply supported. Determine the maximum stress and deflection and compare the results with simple beam theory. Based on the finite element results, explain any violations of beam theory assumptions. The I-beam has a cross section area of 19.11 in.^{2}, a depth of 12.12 in., a flange width of 12 in., a flange thickness of 0.606 in., a web thickness of 0.390 in., and a moment of inertia of 533.4 in.^{4}.

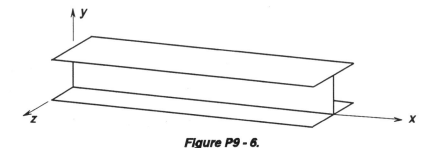

Figure P9 - 6.

9.7   Discontinuity stresses occur when a cylindrical pressure vessel has a hemispherical dome cap such as that in Figure P9-7. Determine the discontinuity stresses in the configuration shown. Run comparative analyses using axisymmetric solid and axisymmetric shell elements. Alternately, you may use a symmetric wedge or quadrant of regular shell elements if the axisymmetric shell element is not available. There is an analytical solution for this problem in many advanced mechanics of materials texts such as [9.11].   What design changes do you envision that will reduce the stress discontinuity?

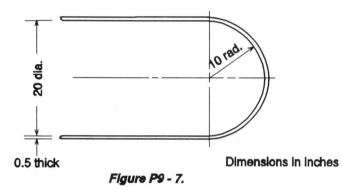

Figure P9 - 7.

9.8   The axle housing in Figure P9-8 is under vertical and side loads from the wheel which transmit through the bearing from the axle. Failures have been occuring where the cylindrical section flares out at a 30° angle. The material is cast steel that is 0.250 in. thick, and the outside diameter of the cylinder is 3 in. You may assume for the finite element model that the flared section is clamped when the diameter reaches 6 in.   Do the stresses indicate a potential for failure? If so, what should be done with the design? If not, what could be causing the failures?

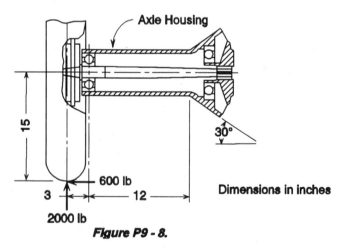

Figure P9 - 8.

9.9   A 20 x 4 mm (OD and wall thickness) steel tube makes a right angle joint with a 50 x 5 mm steel tube using a 6 mm fillet weld with full depth penetration as in Figure P9-9. They have an operating pressure of 10 Mpa. Determine the stresses from a finite element model and select a suitable steel for the application.

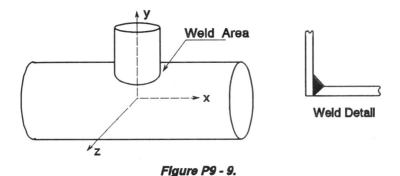

Figure P9 - 9.

References

9.1 Hrabok, M. M. and Hrudey, T. M., "A Review and Catalogue of Plate Bending Finite Elements," Computers and Structures, Vol. 19, No. 3, pp. 479-495, 1984.

9.2 Bathe, K. J., "Plate Bending Elements," in *Finite Element Handbook*, H. Kardestuncer, Editor, McGraw-Hill, New York, 1987.

9.3 Kardestuncer, H., "General Shell Elements," in *Finite Element Handbook*, H. Kardestuncer, Editor, McGraw-Hill, New York, 1987.

9.4 Boresi, A. P. and Sidebottom, O. M., *Advanced Mechanics of Materials*, John Wiley and Sons, New York, 1985.

9.5 Cook, R. D., Malkus, D. S., and Plesha, M. E., *Concepts and Applications of Finite Element Analysis*, Third Edition, John Wiley and Sons, 1989.

9.6 Hughes, T. J. R., *The Finite Element Method*, Prentice-Hall, Englewood Cliffs, New Jersey, 1987.

9.7 Irons, B. and Ahmad, S., *Techniques of Finite Elements*, Ellis Horwood Limited Publishers, West Sussex, England, 1980.

9.8 MacNeal, R. H. and Harder, R. L., "A Proposed Standard Set of Problems to Test Finite Element Accuracy," Finite Elements in Analysis and Design, Vol. 1, No. 1, pp. 3-20, 1985.

9.9 Murthy, S. S. and Gallagher, R. H., "Patch Test Verification of a Triangular Thin-Shell Element Based on Discrete Kirchhoff Theory," Communications in Applied Numerical Methods, Vol. 3, No. 2, pp.83-88, 1987.

9.10 Gupta, A. K. and Ma, P. S., "Error in Eccentric Beam Formulation," Int. J. Numerical Methods in Engineering, Vol. 11, No. 9, pp. 1473-1477, 1977.

9.11 Timoshenko, S., *Strength of Materials, Part II Advanced Theory and Problems*, D. Van Nostrand Company, Inc., Princeton, N. J., 1956.

C H A P T E R 10

DYNAMIC ANALYSIS

When a structure has a load that varies with time, its corresponding response will also vary with time. Up to this point we have discussed only cases in which the load was static on the structure, and therefore the response was static and proportional to the structure stiffness and applied loads. Even for time varying loads, as long as the frequency of the applied loading is less than about one-third of the lowest natural frequency of the structure, the response is given by the static solution in proportion to the instantaneous load. In this case we only need to perform static analysis for the peak set of load conditions expected or the peak combination of loads at various points in time.

However, when the applied loading varies more rapidly, we must employ different solution techniques to include the inertial effects due to material mass and damping effects. There are several different procedures involved in doing dynamic analysis that are dependent upon the type of solution the user is seeking.

This chapter presents the types of dynamic analyses that can be done and the factors involved in performing a finite element analysis for the dynamic problem. The element structural formulations and structure classifications that we have already covered remain valid for the dynamic case. Only the additions to the element formulations required to perform the types of dynamic analyses need presentation.

10.1   Types of Analyses

In the most general case, the problem to solve is equation (10.1) [10.1]. It gives the time dependent response of every node point in the structure by inclusion of equivalent inertial "forces" and damping "forces" in the equation. The inertial forces are given by the product of mass times acceleration, and the damping forces are given by the product of damping coefficient times velocity. The general equation is

$$[M]\{\ddot{D}\} + [C]\{\dot{D}\} + [K]\{D\} = \{F\} \qquad (10.1)$$

where, in matrix form $[M]$ represents the structure mass matrix, $\{\ddot{D}\}$ is the node acceleration vector, $[C]$ is the structure damping matrix, $\{\dot{D}\}$ is the node velocity vector, $[K]$ is the structure stiffness matrix, $\{D\}$ is the node displacement vector, and $\{F\}$ is the applied time varying nodal load vector.

This equation is the set of differential equations of motion in matrix form for the dynamic response of any given structure modeled with a finite number of degrees-of-freedom. While the finite element formulation for static geometrical response follows an integral approach rather than a differential equation approach, most of the time dimension formulations of the numerical approximation follow the differential form. Solution of this equation produces the time varying response of the structure to any specified set of load inputs.

However, the solution to this set of equations taken incrementally in time involves many, maybe even thousands, of static solutions to generate the complete time history of the response. This obviously becomes impractical for any significant length of time cycle. Typically in engineering applications we have more interest in response of the structure to specific types of inputs and the vibrational character of the structure as it relates to these inputs.

10.1.1   Eigenvalue Analysis

The most common type of dynamic analysis for structures is the natural frequency or *eigenvalue analysis*. In many engineering applications we are interested in the values of the natural frequencies of vibration for a structure. The corresponding mode shapes of vibration as the structure deforms in response to inputs at its natural frequencies are also of interest. These are the undamped free vibration response of the structure caused by an initial disturbance from the static equilibrium position.

This eigenvalue problem [10.1] derives from equation (10.1) after zeroing the damping coefficients and applied forces. The structure vibration starts by an initial condition of displacement, velocity or acceleration. Next assume that the motion of every node of the finite element model is a sinusoidal function of the peak displacement amplitude for that node. Substitute for the acceleration components as follows. Define the displacement vector in equation (10.2), where, $\{A\}$ is the vector of peak

$${D} = {A}\sin(\omega t) \qquad\qquad (10.2)$$

displacements for every node displacement component in the finite element model, and ω is the circular frequency of vibration. The velocity vector is then

$${\dot{D}} = {A}\omega \cos(\omega t) \qquad\qquad (10.3)$$

and the acceleration is

$${\ddot{D}} = -{A}\omega^2 \sin(\omega t) \ . \qquad\qquad (10.4)$$

Substituting these into the general equation produces the eigenvalue equation

$$([K] - \lambda[M]){A} = {0} \qquad\qquad (10.5)$$

where, the eigenvalue, λ, is equal to ω^2, and ${A}$ is the eigenvector of peak node displacements associated with each specific value of λ commonly called the mode shape.

There are the same number of independent eigenvalues as there are total degrees-of-freedom in the finite element model. Each of the eigenvalues has an independent eigenvector or mode shape. Since the eigenvectors will not be null vectors, the equation to solve for the eigenvalues will be

$$[K] - \lambda[M] = {0} \ . \qquad\qquad (10.6)$$

After finding the eigenvalues and natural frequencies in the structure, find the corresponding mode shapes by substitution in equation (10.5). The mode shape vector is a set of relative node displacements usually normalized with respect to the maximum displacement component in the vector. The input conditions that initiate the vibration control the actual amplitudes of vibration in any given problem.

So the total solution results in one natural frequency for each DOF in the finite element model. However, we usually only need a few of the lowest eigenvalues of the given model. In fact, in any finite element approximation of the structure, the higher eigenvalues and eigenvectors are inaccurate.

Theoretically the solution for the eigenvalue equation implies that if we deform any structure into its first mode shape, for example, when released, it would continue to vibrate in that mode shape indefinitely. Practically, however, there is always some damping present in any mechanical system, and therefore the vibrations eventually decay.

10.1.2   Frequency Response Analysis

Another type of dynamic analysis of interest is the steady state response of the structure to a harmonic force input at a given frequency. This response may be needed for a range of frequencies. In the frequency response analysis the displacement response of the structure to a harmonic input is also harmonic and occurs at the same frequency [10.2]. Define the force inputs by the equation

$$\{F\} = \{F_0\}e^{i\omega t} \tag{10.7}$$

where, each node force component is given by its peak amplitude, $\{F_0\}$, with a harmonic variation at frequency, ω, in the complex domain. The corresponding node displacements will take the form

$$\{D\} = \{D_0\}e^{i\omega t} \tag{10.8}$$

then the velocity is

$$\{\dot{D}\} = \{D_0\}i\omega e^{i\omega t} \tag{10.9}$$

and the acceleration is

$$\{\ddot{D}\} = -\{D_0\}\omega^2 e^{i\omega t} . \tag{10.10}$$

The governing equation for this frequency response analysis may be derived by substitution of these quantities into equation (10.1). That action results in

$$(-\omega^2[M] + i\omega[C] + [K])\{D_0\} = \{F_0\} \tag{10.11}$$

where the solution for the displacement amplitudes, $\{D_0\}$, is clearly a function of frequency, damping, and force amplitudes, $\{F_0\}$. Solving this equation over a range of discrete frequency inputs determines the vibration frequency response of the structure.

The displacement amplitudes in this case define the deformed structural shape. It is generally not the same shape as the mode shape of the structure at a natural frequency unless the frequency of input happens to coincide with a natural frequency. Most structures are lightly damped and thus we may neglect the damping to simplify the solution in many cases. If we neglect damping in this equation then we may compute the responses for all frequency values except for those which equal a natural frequency. An input frequency equal to a natural frequency produces an infinite displacement response when there is no damping, and the numerical solution algorithm will fail. This is not a serious problem because we can calculate the response near the natural frequency, and the response right on the natural frequency is highly dependent on the amount of damping present anyway. Determining the amount of damping is a very difficult process so the calculated response will not be very reliable.

10.1.3   Transient Response Analysis

If the input loading function is not harmonic, but an arbitrary time dependent function, then we must perform a *transient response analysis*. There are two general approaches to solving the transient dynamic response problem [10.1]. One of these is direct integration of the system equations after approximation by a finite difference or finite element method in the time dimension for the velocity and acceleration components. This direct integration approach will obviously involve the total set of system equations and must perform many time steps with a complete solution in each step. This can become a large computing task for significant size problems.

The second approach called *modal superposition* helps lessen part of this computing problem. The basis of this approach is an assumption that superposition of the mode shapes corresponding to the lower natural frequencies adequately represents the dynamic response of the structure. The complete response is found by the summation of correct fractions of the low frequency mode shapes. Mathematically this amounts to a transformation of the equations from node displacement coordinates to a set of modal coordinates. The transformation changes the set of system equations consisting of one equation for each DOF in the model to a set of modal equations involving the selected number of mode shapes. This results in much fewer equations. This is also an approximation of the total structural response, but in most cases of structural vibration response it has shown to be sufficiently accurate.

Both methods have several different algorithms to use for the solution. Further discussion here is beyond the scope of this text, but much more detailed discussions are available in many advanced finite element texts.

In any of these types of analyses we can see that the complexity of the problem in finding the dynamic response of the structure is much higher than for the static response. These complexities translate to greater difficulty in achieving a reliable dynamic solution.

Part of the difficulty with dynamic analysis is that structures that have vibration problems are usually thin, and we have to model them with plate or shell elements. These elements are already known to have serious drawbacks in static analysis.

10.2 Formulating Mass and Damping Matrices

These solution procedures need to have a structure mass matrix and structure damping matrix to include damping. As in the formulation for the structure stiffness matrix, this is done by setting up an element level matrix first and then making the proper assembly for the structure. The element mass matrix develops by employing the D'Alembert principle. The inertial effects become body forces produced by taking the product of the acceleration field over the element with the mass density. These body forces convert to nodal loads through the principle of virtual work utilized in the element stiffness matrix formulation [10.3]. The resulting expression is

$$\{f\} = \int_V [N]^T \{g\} dV \qquad (10.12)$$

where, $\{f\}$ are the element nodal forces, $[N]$ are the interpolation functions, and $\{g\}$ are the body forces. Replace $\{g\}$ with

$$\{g\} = -\rho \{\ddot{u}\} \qquad (10.13)$$

where, ρ is the mass density, and $\{\ddot{u}\}$ is the acceleration of the differential volume, dV, within the element field.

Using the same set of interpolation functions for acceleration that are used for displacement gives

$$\{\ddot{u}\} = [N]\{\ddot{d}\} \qquad (10.14)$$

then compute the element nodal loads by

$$\{f\} = \left(\rho \int_V [N]^T [N] dV\right) \{\ddot{d}\} . \qquad (10.15)$$

Then from this form we see that the element mass matrix is given by

$$\{m\} = \left(\rho \int_V [N]^T [N] dV \right) . \tag{10.16}$$

This element mass matrix is called the *consistent mass matrix* because the formulation uses the same interpolation functions used to formulate the stiffness matrix.

The assembly of all the element mass matrices into the structure mass matrix follows the same conceptual approach used to assemble the structure stiffness matrix. Because the acceleration field converted to a field of body forces, the resulting forces then must satisfy the equilibrium equations.

In principle, we could follow the same approach to develop the structure damping matrix. If damping was truly a viscous function that yields damping forces only proportional to point velocities, then this would be the correct approach and yield the most accurate representation. In fact, damping is not a very well understood mechanism. It is known that damping also relates to displacement amplitude and accelerations as well through coulomb friction, and other energy loss mechanisms. These effects in application then would produce coefficients on the displacement vector as well as the acceleration vector so only applying damping on the velocity vector is already a major simplification of the actual system. This is obviously very difficult to understand and make a reasonable approximation in most cases.

Fortunately, in most structural vibration problems damping is fairly small, so that the approximation usually works. Proportional damping is a technique we use commonly, and it seems to work reasonably well for most structural problems. In this approach the structural damping matrix forms from a linear combination of fractions of the structure stiffness and the mass matrix [10.1] as given by equation (10.17).

$$[C] = \alpha[K] + \beta[M] \tag{10.17}$$

The user or analyst is then responsible for selecting the values of α and β. This is usually done by collection of experimental measurements on a similar material and structure. This approach simplifies the equations of motion by uncoupling them into a set of simultaneous ordinary differential equations so that the transient response analysis by direct integration or by modal superposition becomes a much easier task.

Although the formulation for the mass matrix seems conceptually straightforward for the finite element formulation, it results in an element mass matrix which is full and symmetric. The assembly of all the element matrices results in a banded structure mass matrix that is similar to the stiffness matrix. In some cases the off-diagonal terms of the mass matrix generate numerical difficulties in the solution procedures.

Another type of mass matrix formulation is a *lumped mass matrix* [10.4]. In this case the material mass is lumped into point masses at the node points, and therefore they have only translational inertia with no rotary inertia for each lumped mass. The distribution of particles accounts for the rotary inertia of the body. This concept produces a diagonal mass matrix, and this diagonal matrix significantly reduces the numerical problems associated with the off-diagonal terms of the consistent mass matrix. It also simplifies some of the transient response analysis procedures.

There is no exact prescription for producing the lumped mass matrix. It may be done simply by distributing the mass for each element equally to the nodes of the element and then coalescing the contribution from each element at each node. It may also be done by computing the consistent mass matrix, discarding the off-diagonal terms, and scaling the diagonal terms so that the sum of diagonal terms equals the total mass [10.4].

It is not clear that either the consistent or lumped mass matrix is superior for all problems. One works better for some problems and the other works better for different problems. Usually the consistent mass matrix performs better for coarser meshes or models because it includes the rotary inertia terms at the nodes. The lumped mass approach usually works better for models with more refined meshes since the rotary inertia terms are much smaller and the mass matrix is much cheaper and quicker to assemble.

There are numerical difficulties associated with a large set of equations in any of these dynamic analyses. An approach called "static condensation" or "Guyan reduction" [10.5] is commonly used in many commercial finite element programs to reduce the number of equations. In this approach we first develop a model to represent the geometry and expected deformation shapes of the structure adequately. Then from the total degrees-of-freedom in the model, we select a group of master DOF on the basis that they are the more dominant DOF in the model in representing the structural response. The term "dominant" means that the actions of these master node DOF control the vibration and the remaining DOF of the model become "slaves" because they must follow the pattern defined by the response of the master nodes.

We omit the mathematical development here, but we will discuss some of the basic features of this approach [10.4]. Some programs expect the user to select the masters, and the program then accepts a list and proceeds to do the numerical manipulation. The numerical algorithm partitions the set of system equations to group the master DOF and then condenses the slave DOF from the set of equations. The condensation is done by discarding the mass associated with slave nodes and then setting up the displacement relations between master and slave DOF which are then only a function of the structure stiffness components relating these DOF. This produces a transformation of the mass and stiffness terms associated with master DOF so that we have a reduced set of equations to solve only involving the master DOF which we may then solve using one of the dynamic analysis procedures. Using the solution for displacements for the master DOF, we can then back substitute and produce the displacements for all the slave DOF and yield a good representation of the deformed shape geometry.

The selection of master DOF usually points to the DOF which dominate the lower frequency response since most vibration problems occur in the lower frequency modes. Also, since we neglect the mass effects associated with slave nodes in the condensation step, the master DOF must be those with larger fractions of structure mass. Those DOF then have a higher ratio of mass-to-stiffness. Most commercial programs have an automatic selection feature of master nodes based on this premise and the user must then only specify how many master DOF to use. The number of masters then becomes a parameter for us to vary until we reach sufficient accuracy of the solution. We should be sure to distribute the master nodes throughout the model without clusters. The high mass-to-stiffness ratio requirement for masters mostly eliminates the rotational DOF in beam, plate or shell elements from consideration.

An example of selection of master DOF is shown in Figure 10-1. A cantilevered thin plate model uses plate elements having five DOF per node. Imagine that the mass at each node represents a point mass equal to the mass of material around the node and the rotary inertia for the same material block. The point mass relates to the translation DOF and the rotary inertia relates to the rotation DOF.

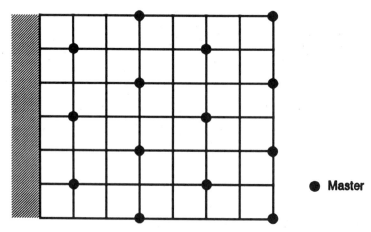

● Master

Figure 10-1. Master DOF Selection for Analysis of a Cantilevered Plate

First, consider that for a refined mesh the rotary inertia effects will be much less than the point mass inertia effects. This leads to discarding of the rotation DOF at the nodes. Then for a given point mass at the node, the stiffness for inplane translation will be much higher than for the lateral translation DOF. Therefore, discarding all the DOF except for out-of-plane displacement reduces the equation set by a factor of five. At this stage there would be one equation per node to solve for the eigenvalues of the plate. Further selection of masters would discard DOF uniformly across the mesh and at the fixed edge as sketched in the figure.

10.3   The Finite Element Model

The finite element model must be compatible with the type of analysis to be done. The first step is to choose the proper element to use for dynamic modeling, and it is not always the same element you would use for the static analysis. For example, if the static case is 2-D plane stress with a small thickness, then the lowest natural frequencies of vibration and corresponding mode shapes may involve out-of-plane motion. So for a model we analyze in 2-D plane stress for the static loading, we may need to model with a plate or shell element for an eigenvalue analysis to determine all the natural frequencies and mode shapes.

You should know that the choice of element may eliminate part of the solution. If a 2-D plane stress element is chosen for a dynamic model, then we will find only the natural frequencies and mode shapes involving inplane motion. If a plate element having no membrane (in-plane) DOF is chosen for a model, then we only get flexural modes of vibration since the element formulation excludes any in-plane motions. If we use a 2-D beam element for a model, then we get only the natural frequencies and mode shapes lying in the 2-D plane. However, the natural frequencies and mode shapes involving out-of-plane motion might be lower and excited first in actual service of the beam structure, but the finite element model would not have predicted their existence.

The mesh requirements for the dynamic case also depend on the type of problem and the output of importance. If we are looking for only the natural frequencies, then we can use a relatively coarse mesh, while accurate mode shapes need a more refined mesh. To get stresses resulting from the vibration modes requires an even more refined mesh to achieve reasonable accuracy. Also, the mesh plan must provide a reasonable approximation of all mode shapes of interest. If only the first few natural frequencies are important then a relatively coarser mesh may be satisfactory. More mesh refinement accompanies the need for more natural frequencies.

Utilizing symmetry in dynamic models is somewhat tricky. Mode shapes at natural frequencies tend to develop in both symmetric and antisymmetric patterns along with periodic multiples or subdivisions of these patterns. Then we cannot exploit symmetry to reduce the model size without potentially missing some of the natural frequencies of the structure in the model. When using a symmetric model we must specify carefully the displacement boundary conditions on nodes on the planes of symmetry to avoid losing information. Usually more than one case set of boundary conditions must be run to get the total solution.

For example, modeling a circular plate as shown in Figure 10-2 we would expect that the mode shapes for corresponding natural frequencies of that plate to exist in both symmetric and antisymmetric patterns. So if we take half the plate for a model, then apply displacement boundary conditions that force the structure to be symmetric about the line of symmetry, then we find only the natural frequencies which produce mode shapes symmetric about that line. Then we must follow that case with a case where we restrain the nodes on the plane of symmetry to only allow unsymmetric motion. The solution in this case will provide all the natural frequencies and mode shapes for the antisymmetric modes. The combination of the two cases gives the total set of natural frequencies and mode shapes.

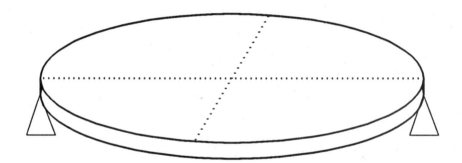

Figure 10-2. A Circular Plate With Support Across a Diameter

Using a quarter section of the plate requires running four cases. These would be: 1) symmetric conditions along both radial edges, 2,3) the symmetric condition on one edge and the antisymmetric condition on the other, and 4) antisymmetric conditions on both edges. Any further symmetry reductions would not allow determination of all the natural frequencies and mode shapes of the structure.

Since we cannot use symmetry as effectively for dynamic models as for static models, then in general, this makes the dynamic solution more difficult to obtain accurately.

We must incorporate any support boundary conditions that actually exist on a structure into the model since the restraint conditions strongly influence the natural frequencies and mode shapes. For example, a beam of given length will have one set of natural frequencies if it is a cantilevered beam and will have a different set if it is a simply supported beam.

In an eigenvalue analysis for natural frequencies, there is no load distribution applied. A frequency response analysis must have the load amplitudes and distribution input along with the frequency. In a dynamic response or time history analysis the load inputs by forces, displacements, velocities, or accelerations are a function of time. Many commercial programs have programmed in the ability to apply an existing design load spectrum such as for simulating earthquake conditions for example.

10.4   The Analysis Step

The analyst is responsible for selecting the type of analysis to perform. In most cases an eigenvalue analysis will be done first, even if another type of analysis is to be done later. The eigenvalue analysis provides valuable insight into the expected behavior of the dynamic structure. An eigenvalue analysis may be done by several different algorithms [10.1]. However, any given program may have only one or two options from which the analyst may choose. In a "static condensation" or "Guyan reduction" process the program will usually automatically select the master DOF, but the analyst must decide how many masters to choose. There should obviously be more masters than the number of eigenvalues that we wish to determine with some degree of accuracy. Usually five times the number of eigenvalues will provide a reasonable starting number. Additional solutions with incrementally higher factors will show convergence of the desired number of eigenvalues.

If the program does not have a condensation process, then the model should be somewhat coarser than would be necessary for stress analysis to assure a reasonable numerical accuracy from the eigenvalue numerical algorithm.

For other types of analyses, the mesh should be able to represent accurately the vibration frequencies up to a value at least three times higher than the highest frequency content of the dynamic applied load [10.1].

There are many different numerical algorithms for eigenvalue analysis, and also many different numerical approaches for the transient response calculation. Explanation of these approaches is beyond the scope of this book. However, the method used by any given program is usually (but not always) adequately referenced so that the user may explore it in greater detail. It is important for the user to have some reasonable understanding of the approach to use and what the potential pitfalls may be, as well as what the potential accuracy may be. Beyond that the user must experiment within the parameters that are user selectable for any given algorithm to achieve the desired results. Of course, experience with prior solutions and correlation or verification of those solutions with other analytical or experimental data is invaluable.

10.5   Output Processing and Evaluation

Results from an eigenvalue analysis will have the requested natural frequencies either from specification of the number of lowest frequencies to find or specification of a frequency range in which to find all the natural frequencies. Node displacement sets for each mode describe the corresponding mode shapes. Normalizing of the mode shapes may be done by setting the maximum amplitude value in each shape to unity and scaling all other values to less than one. The actual displacement amplitude response of a structure excited at one of its natural frequencies theoretically becomes infinite. Therefore, a specific amplitude for the displacement in an eigenvalue analysis is only relative to other values.

Except in very simple cases we cannot picture the mode shape well from examination of printed values. Therefore, the deformed shape plots from an eigenvalue analysis will display the mode shapes. The analyst may specify in the analysis step the number of natural frequencies and corresponding mode shapes to find. The graphic data file of mode shapes will consist of a displacement shape for each of the natural frequencies. So there is a deformed shape plot for each mode shape. We must evaluate each deformed shape then for its ability to depict the shape accurately from the mesh we are using.

If a dynamic response analysis using modal superposition is done following the eigenvalue analysis, then we should check the solution convergence to determine if we have used an adequate number of accurate eigenvalues. For a dynamic response analysis using modal superposition, examine the frequency content of the loading to determine the highest frequency of any significant input loading. Then for an accurate dynamic response analysis by modal superposition, use a mesh which accurately represents all the modes up through a frequency about four times the highest frequency of load input. This mesh requirement also applies to the dynamic response analysis using direct integration since we need to represent a calculated response of the model to these frequencies of load input accurately.

Continue mesh refinement until you get sufficient accuracy of the important eigenvalues. The meshes for dynamic models are normally nearly regular meshes. Examine deformed shapes of modes of interest, especially those at higher frequencies, very carefully to verify a smooth representation of the displacement field. Refine the mesh in those areas where the displacement field is varying most rapidly and therefore needs the refined mesh to make a smoother fit.

Stress contour plots for each mode shape from an eigenvalue analysis also may be available, but keep in mind that the magnitudes are scaled to the normalized deformed shape. This may make the values extremely high and in any case only represent relative values because the actual displacement amplitudes are not known. However, stress contour plots from a transient response analysis should represent actual values if we have accurate displacements.

10.6   Case Studies

The case studies will begin with a simple tapered cantilevered beam for which we will find the first five eigenvalues and mode shapes. The steel beam is 10 in. long, 0.100 in. thick, and its width varies from 2.0 in. at the wall to 1.8 in. at the free end [10.6]. We will use several models to demonstrate different modeling approaches and mesh convergence. First beam element models are used then plate element models later.

A five element beam model produced the results in Figure 10-3. The beam orientation has the x axis along the 10 in. length, the y axis across the 2 in. width, and the z axis through the 0.100 in. thickness. The first mode is clearly z motion of the beam end, and the second mode is the next z motion mode. The third mode is torsional about the x axis, and graphically is simply a straight line the length of the beam. The fourth mode is the third z motion mode, and the fifth mode is the first y motion mode. The coarseness of the mesh is very apparent in the mode shape plots.

Changing to a twenty element beam model produced the results in Figure 10-4. The values of the frequencies differ by less than 1 percent except the torsional mode which has only 2 percent difference. The mode shape plots are significantly smoother, but do not show any real difference in character except for the slope constraint at the wall.

A plate element model with a two by five element mesh gives the results in Figure 10-5. These values of frequency are within 2 percent of the beam element models except the torsional mode 3 which is about 8 percent higher, and can be graphically illustrated by the plate model.

Refining the model with a four by ten element mesh results in Figure 10-6. This refinement only improved the frequency values by about 1 percent, with some improvement in visual quality.

This study shows that we can get reasonably accurate values of natural frequency and mode shape for this structure using relatively coarse models of either beam or plate elements.   This is also usually true for finding eigenvalues and mode shapes for most structures.   Coarse models usually provide good values for the lower eigenvalues that are the ones of most importance.

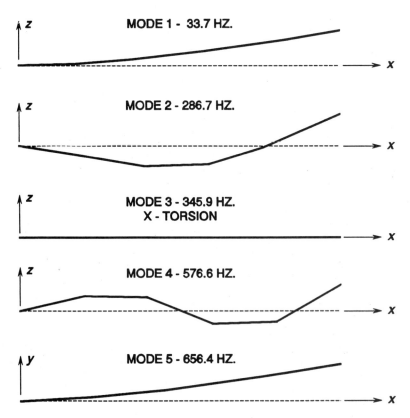

Figure 10-3. Natural Frequencies and Modes in a Five Element Model

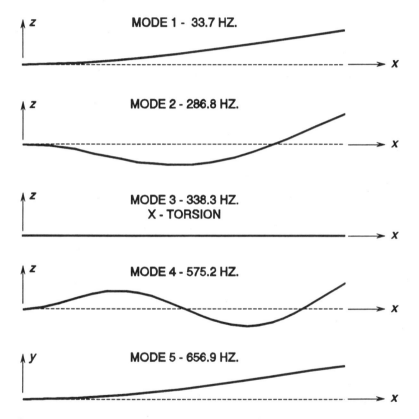

Figure 10-4. Natural Frequencies and Modes in a Twenty Element Model

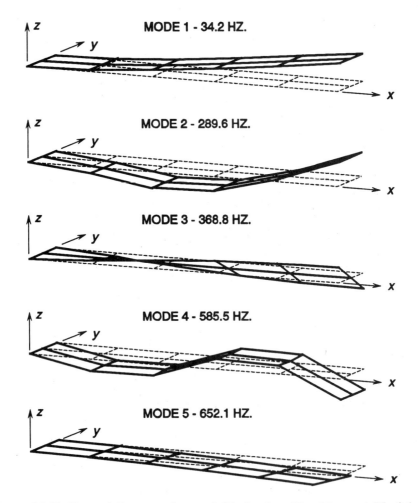

Figure 10-5. Natural Frequencies and Modes in a Ten Element Model

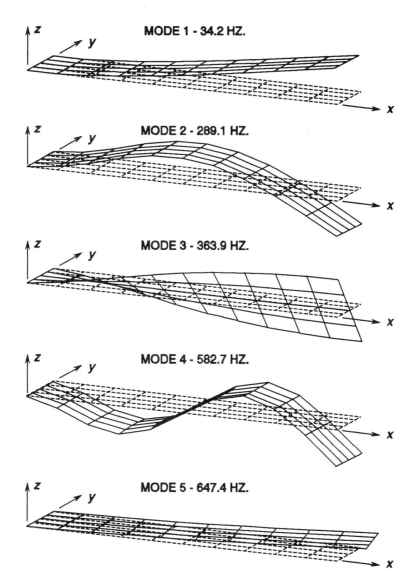

Figure 10-6. Natural Frequencies and Modes in a Forty Element Model

   The next case will examine the transient response of a cantilevered
plate. The steel plate is 2 in. long, 1 in. wide and 0.100 in. thick. The first
step in any transient analysis is the eigenvalue solution. The first four
natural frequencies and modes shapes are shown in Figure 10-7. The first
mode is z motion typical of cantilevered beams. The second mode is the
first torsion motion about the x axis. The third mode is the second z
motion mode, and the fourth mode is the second torsional mode. In this
analysis we employ static condensation described in Section 10.2 to use
only the z translation and x and y rotation of the unrestrained nodes as
master DOF.

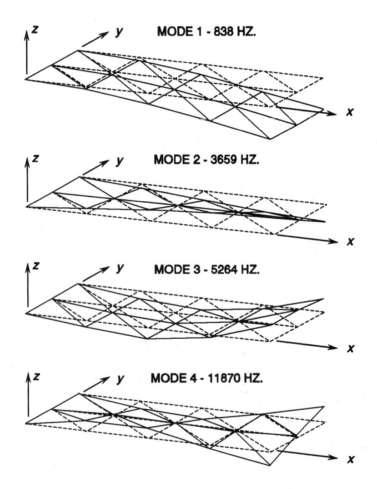

Figure 10-7. Four Modes of a 2 by 1 by 0.100 Cantilevered Plate

   Two transient analyses will be done. In each case the nearest corner in
view will have a specified displacement input. To control the value of

displacement, the input device must be very rigid, and therefore will act as an additional fixed restraint at the corner. This changes the eigenvalues and mode shapes. The first four frequencies are now 2083, 4698, 8829, and 13580 hertz.

It is obvious this restraint stiffens the structure a great deal. If the input were a forcing function with very low stiffness, the plate eigenvalues would not change. The first analysis is for input of a step displacement of 0.01 in. held for 0.44 ms and then ramped down over a 1.5 ms interval. The solution method was modal superposition using the first four modes. The response of the other corner of the plate is shown in Figure 10-8. The curve labeled A is the input displacement of the near corner, and the curve labeled B is the response of the far corner.

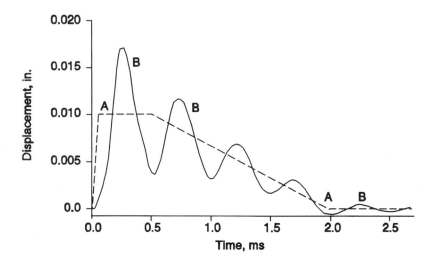

Figure 10-8. Transient Response of the Plate to Specified Displacement

The next analysis has a harmonic displacement input at the near corner of amplitude 0.01 in. over a frequency range of 1000 to 6000 hertz. The displacement frequency response of the far corner is given in Figure 10-9. The curve labeled B is the constant amplitude input displacement. The curve labeled A is the amplitude vibration response of the far corner at the given frequency. The amplitude obviously increases rapidly near the first eigenvalue at 2083 hertz, and the second eigenvalue at 4698 hertz which are within the input frequency range. The displacement shape at 2050 hertz as the frequency approaches the first natural frequency is shown in Figure 10-10.

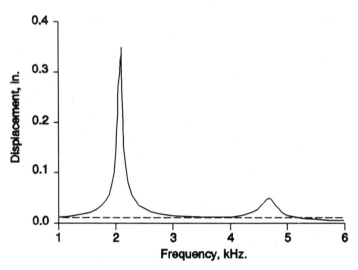

Figure 10-9. Displacement Frequency Response of the Plate

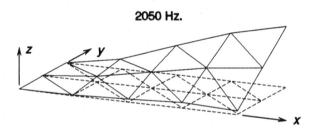

Figure 10-10. Displacement Shape at a Frequency of 2050 Hertz

Finally the stress level over the frequency range is shown in Figure 10-11. This is the value of the Von Mises equivalent stress in the corner element where the input displacement applies. The stress scale was set to 100 kpsi, but the maximum amplitude reaches 486 kpsi near the first resonant frequency. The maximum amplitude near the second natural frequency is 54 kpsi. The two curves here represent the real and imaginary components of the stress. The real component is in solid line and the imaginary in dashed line. As the driving frequency goes through resonance the response phase shifts 180 degrees causing the real component to dip and the imaginary component to peak as shown in the plots. The calculations were done at 10 hertz intervals.

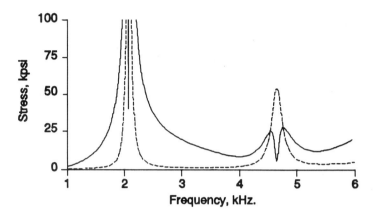

Figure 10-11. Stress Frequency Response of the Plate

Problems

10.1 Recall the framework structure of Problem 3.6 in Figure P10-1. Determine the first four natural frequencies and mode shapes in the plane of the given structure, first using truss elements and then using beam elements. The steel members are 2-in. square tubing with ⅛-in. wall thickness. Replace the 12,000 lb. and vertical component of the 15,000 lb. load by steel masses.

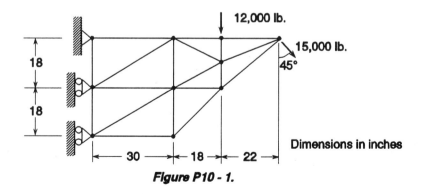

Figure P10 - 1.

10.2 Recall the framework structure of Problem 3.7 in Figure P10-2. Determine the first four natural frequencies and mode shapes in the plane of the given structure, first using truss elements and then using beam elements. The steel members are 2-in. square tubing with ⅛-in. wall thickness. Replace the vertical component of the 30,000 lb. load by a steel mass.

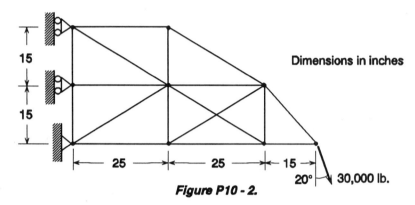

Figure P10 - 2.

10.3 A rectangular plate of 0.125 in. thick stainless steel is shown in Figure P10-3. Find the first four natural frequencies and mode shapes for all edges free, all edges simply supported, and all edges fixed.

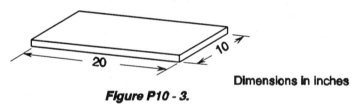

Dimensions in inches

Figure P10 - 3.

10.4 Recall the L-shaped plate of Problem 9.1 in Figure P10-4. The material is 0.100 in. aluminum. Find the first four natural frequencies and mode shapes. Compare with a solution for a fixed ends beam of length equal to the sum of the midline lengths of the two legs on this plate. Neglect any mass associated with the 20 lb. force.

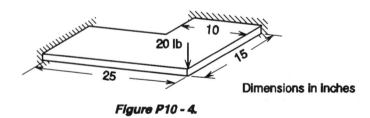

Dimensions in inches

Figure P10 - 4.

10.5 Recall the compressor flap valve from Problem 9.3 in figure P10-5. It is 0.018 in. thick steel. Find the first three natural frequencies and mode shapes if the valve is simply supported under the ears at a spacing equal to the outer diameter.

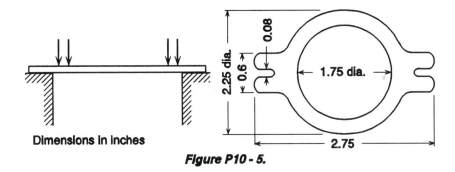

Figure P10 - 5.

10.6 The route for an exhaust tube for a refrigeration compressor is shown in Figure P10-6. It is 0.25 in. ID, 0.060 in. wall thickness steel tubing. In the case when the lower end is welded into a comparatively rigid housing and the upper end is free, determine the five lowest eigenvalues and corresponding mode vectors.

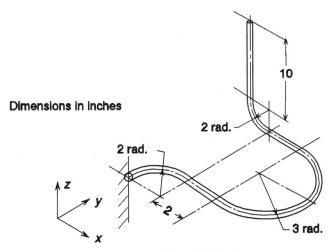

Figure P10 - 6.

10.7 Find the first five natural frequencies and mode shapes of the compression coil spring sketched in Figure P10-7. The material is A228 Music wire. The spring has an inside diameter of 1.100 in., a wire diameter of 0.148 in., and a free length of 1.575 in. There are

5½ total coils but only 3½ active coils with the ends being closed and ground. Assume the ends are fixed.

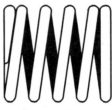

Figure P10 - 7.

10.8 The steel blade in Figure P10-8. operates at rotation speeds up to 10,000 rpm. Find its three lowest natural frequencies and mode shapes. Are there additional natural frequencies in the operating speed range? Would you expect any operational problems with this device?

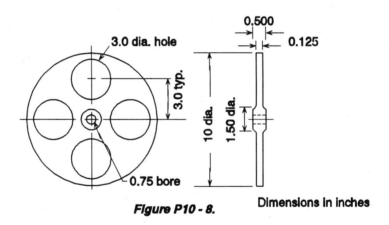

Figure P10 - 8.   Dimensions in inches

References

10.1 Bathe, K. J., *Finite Element Procedures in Engineering Analysis*, Prentice-Hall, Inc., Englewood Cliffs, New Jersey, 1982.

10.2 Zienkiewicz, O. C., *The Finite Element Method*, Third Edition, McGraw-Hill, London, 1977.

10.3 Fenner, D. N., *Engineering Stress Analysis, A Finite Element Approach with Fortran 77 Software*, Halsted Press, John Wiley and Sons, 1987.

10.4 Cook, R. D., Malkus, D. S., and Plesha, M. E., *Concepts and Applications of Finite Element Analysis*, Third Edition, John Wiley and Sons, New York, 1989.

10.5 Guyan, R. J., "Reduction of Stiffness and Mass Matrices," AIAA Journal, Vol. 3, No. 2, p. 380, 1965.

10.6 *MSC-PAL2, Advanced Stress and Vibration Analysis*, User's Manual, The MacNeal-Schwendler Corporation, Los Angeles, CA, 1990.

CHAPTER 11

ADVANCED MODELING TOPICS

There are a few modeling techniques that deserve special discussion. These topics do not fit well into any of the earlier chapters, but may be of substantial benefit in developing an accurate representation of the physical problem in the finite element model. In effect these topics apply to all of the earlier chapters, but they are better understood after having covered the material from those chapters.

Application of these principles and techniques will make the finite element model much more "realistic" in relating to the actual behavior. It is very common in mechanical design problems to have several parts to a device that contact and interact with each other. The finite element formulation represents the continuum of material in a single part. Therefore there must be some special procedures to couple two or more parts so they interact properly.

One type of element may represent some components very well overall yet the component may have some local features it does not represent at all. We may address this problem by use of different element types in the same model if we connect the different elements properly. We also may solve this problem by use of a model with the element best suited for the overall problem, and follow the first analysis by a subregion model of the local feature using an appropriate element type with input from the parent model.

Constraint equations provide a way to force a displacement behavior in the model that matches some physical condition through equations that provide kinematic relations between two or more degrees-of-freedom.

Much active research is being conducted on methods to assure accurate finite element models. The process of convergence of the numerical modeling done by finite elements through successive refinement is being automated. Adaptive meshing programs the computer to estimate errors in a given finite element model and refine the mesh in accordance with the magnitude of distributed errors until it reaches an acceptable level of error.

11.1   Multi-Component Interfaces

Most machines and structures consist of more than one part. Therefore when the machine or structure undergoes loading, they get shared or sent through several interfaces between the multiple components. However, stress analysis in general, and the finite element method in particular, apply to a continuum of material making up the component under analysis. Typically, we idealize the interfaces between components or ignore them to perform a stress analysis by such measures as concentrated forces, fixed or hinged supports, uniform pressure or other approximations. This approach usually assumes a static determinacy which may not be true, but does allow considerable simplification of the problem.

This is a problem area where the power of the finite element method can provide a more realistic solution. We can easily construct finite element models of each component and then we must connect them. Connection by making one continuous model of the multiple parts is usually not acceptable. This assumes a perfect bond on the interfaces with no provision for changes in the interface connection or relative motion of the parts on either side of the interface. However, without some connection method, one individual model would not even recognize the existence of another model even if we position them adjacently in space. For example, we might model a wheel resting on a rail as pictured in Figure 11-1 with the symmetric mesh in Figure 11-2. There is one mesh for the wheel and another mesh for the rail. We positioned them so that one of the nodes in each mesh has a coincident location at the point of contact. However, as long as they are separate nodes, when a load applies to the wheel it will not transfer to the rail.

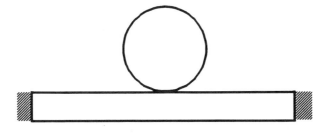

Figure 11-1. Wheel and Rail

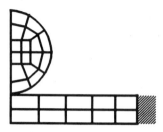

Figure 11-2. Wheel and Rail Symmetric Model

One approach is simply to replace the coincident nodes with a single node that works in each mesh. This will provide point contact representation of the contact area between the two components. As long as the local contact stress distribution is not of interest, then the model will give a reasonable solution.   Of course, in this case it would be the same as modeling each component separately with an applied concentrated load on each at the point of contact.

Actually the local contact stress problem is nonlinear.   As the load gradually increases, the contact area between the two parts increases because the material deforms in a nonlinear relation to the magnitude of applied load. This changes the distribution of load in the contact area and therefore changes the stress distribution.   We can use successive linear elastic solutions with increments in load to solve the nonlinear problem as long as the material remains elastic.

Begin with the two coincident nodes merged and apply load until the deformations in each component bring the next two adjacent nodes on the surface into coincidence. Merge these two nodes and proceed with further loading until the next two nodes are coincident. Continue until you reach the total load magnitude. Of course, accumulate the stress increments as the load increments. Unfortunately, this kind of approach does not allow the surfaces to have any relative slip, so it may not work in general interface conditions.

The need for connection of multicomponent interfaces has led to the development of special elements usually called *gap elements* and other algorithms for allowing surfaces to contact and slide without penetration or material overlap in the models. Gap elements may use truss elements for a basis or thin solid elements with special formulations to avoid the very high stiffness associated with the thin direction [11.1].   Surface algorithms called *slideline* [11.2] or *contact segment* [11.3] algorithms continually check for penetration of one component by the surface of another and push it back.

The simplest gap element uses a truss element with bilinear material properties and is available in several commercial codes [11.4]. The two surfaces illustrated in Figure 11-3 connect through truss gap elements shown by a spring symbol. Although the space between surfaces is large in the illustration, the actual length of the springs may be very small, even zero as long as the alignment of the spring is known.

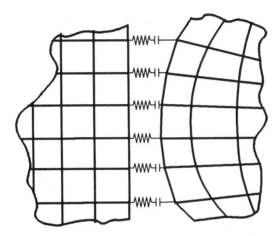

Figure 11-3. Two Surfaces Connected by Truss Gap Elements

Truss elements are one-dimensional in that they only have stiffness against axial motion. In this arrangement they provide stiffness against motion of the surfaces normal to the line of contact. They offer no resistance to motion parallel to the line of contact. Thus, they simulate frictionless contact of the two surfaces. If the surfaces were flat, then contact would exist initially along the whole surface. For curved surfaces contact begins at some point and spreads as the load increases. An initial gap in each element allows for space which must close before the element becomes active.

A compressive force in the gap element maintains contact, but if the force becomes tensile due to load shifts or changes in the contact area, the components should be free to separate. Simulating this action requires the spring stiffness to be bilinear as sketched in Figure 11-4 with a high stiffness in compression and zero stiffness in tension (or a very small value to avoid numerical problems). Addition of lateral stiffness to the element formulation models friction effects along the interface as in Figure 11-5. The lateral stiffness levels off after reaching a shear force to normal force ratio equal to the static coefficient of friction.

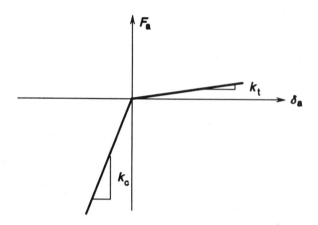

Figure 11-4. Bilinear Normal Stiffness of Truss Gap Elements

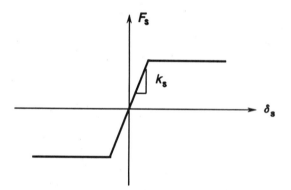

Figure 11-5. Lateral Stiffness of Truss Gap Elements

Two-dimensional area interface elements are shown in the interface of Figure 11-6. Since the thickness is much less than the length of the element along the interface, a standard 2-D element formulation will generate a very high stiffness in the thickness direction. The result is ill-conditioning of the structure stiffness matrix and likely failure of the computer solution. Special formulations based on relative displacements across the thickness or stiffness modification deal with the problem [11.5, 11.6]. These elements also may use material properties simi- lar to those above to account for separation and friction. The same principle extends to 3-D interfaces.

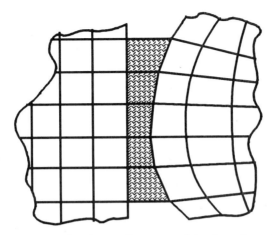

Figure 11-6. Two Surfaces Connected by Interface Elements

Use of gap or interface elements usually requires a matching node pattern and spacing along the surfaces that will contact. Algorithms for slidelines or contact patches usually do not require matching node patterns [11.2]. In these algorithms, once you identify the likely surfaces that will contact, you pick one surface and the computer program locates the points of contact of nodes on the other surface as in Figure 11-7. During each load increment applied to the structure the program checks each node in contact for penetration into the surface. If penetration occurs, it applies enough force normal to the surface to restore the node to a location along the line between the two nodes of the picked surface. Development of 3-D contact algorithms is much more complex and is a current problem in active research.

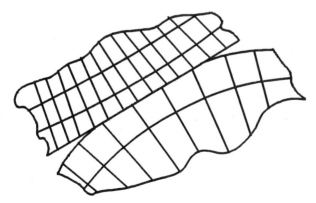

Figure 11-7. Two Surfaces in Slideline Contact

These procedures all provide a method for dealing with the connection of multicomponent interfaces. Remember that these are actually nonlinear problems that we are solving by iteration of linear solutions. Therefore in any given load increment the solution must iterate until the contact configuration is stable before applying the next load increment.

11.2   Mixing Element Types

In many structures the model will be better if made up of more than one element type. We could represent different parts of the structure with an element selected to match its behavior. For example, we could model a simple wrench shown in Figure 11-8 with beam and 2-D plane stress elements as shown in Figure 11-9. However, to do so requires that we properly couple the different element types for the model to be sound [11.1].

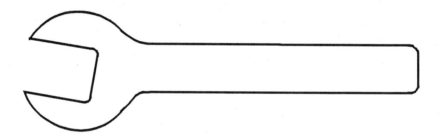

Figure 11-8. A Simple Wrench

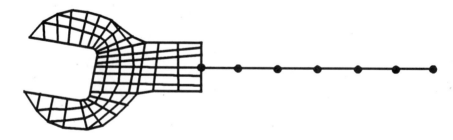

Figure 11-9. A Beam and 2-D Plane Stress Wrench Model

In this example, the connection of the beam element to the 2-D elements must have more coupling than just the common node in the model. A 2-D beam element has three DOF per node, the two in-plane translations and the rotation vector normal to the plane. The plane stress elements only have two DOF per node, the two in-plane translations. Therefore connecting the beam element to the plane 2-D elements by using a common node leaves the beam free to rotate relative to the plane element portion of the model. The use of constraint equations can attach the beam's rotation DOF to the rotation of the edge of the plane element portion. Here, constraint equations would force the beam's rotation DOF to be equal to the rotation of the edge, and the translation normal to the edge would lie in the plane of the beam's rotated cross section. We will discuss constraint equations in the next section of this chapter.

This example illustrates the requirement for matching and connecting nodal DOF when different element types join in a model. However, even proper connection cannot generate more information than is available from either of the element types. Connecting a 2-D truss element to a 2-D beam element will still behave as a pinned joint. This is because the truss element has no rotational DOF, and no kind of constraint equation will change it. If we fix the rotational DOF of the beam at the joint to zero in the model, then there must be some physical source to support that fixity or it is artificial.

Two-dimensional trusses may connect to 2-D plane elements or 3-D trusses to 3-D solid elements to represent support conditions, slender member ties such as wheel spokes, or gap elements as discussed in the previous section. These connections do not require any constraint of DOF since the DOF of each element are the same. Mixing 2-D and 3-D elements which only have translation DOF in the same model produces another condition that we cannot resolve by constraint equations. Only the in-plane response of the 2-D elements can influence the 3-D elements, but this may be acceptable in some models.

Connection of plate or shell elements to 3-D solid elements is similar to the 2-D beam to 2-D plane elements discussed above. It is a little more complex because of the additional DOF involved, but it generally requires coupling the rotational DOF to the corresponding equivalent rotation of a group of nearby nodes. This is shown in Figure 11-10 where shell elements are joined to 3-D solid elements [11.7]. At each node along the joint, both the rotation DOF in the plane tangent to the shell surface must couple to the corresponding rotations of the group of nodes through the thickness of the solid elements. Also, the line of nodes running through the thickness must displace and rotate as a straight line.

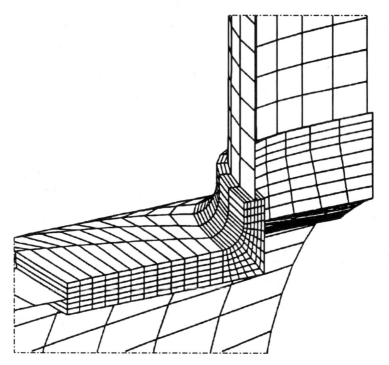

Figure 11-10. A Combined Shell and 3-D Solid Element Model

11.3   Constraint Equations

Constraints provide a method for having a prescribed relation between two or more DOF in a model [11.1]. This may be useful for certain types of displacement loading, for multicomponent interfaces, connecting different types of elements, and other reasons. These relations must satisfy some physical aspect of the real structures or provide some mathematical equivalency in the model for them to be valid.

Typically, the input of these relations in computer programs is through constraint equations. We must formulate equations which provide the desired relation. Refer to the wrench model of Figure 11-9 in the enlarged view of Figure 11-11. We need to constrain the numbered nodes by input of constraint equations.

At the common node 6, both the *u* and *v* displacement components are already the same for the beam element and the 2-D plane elements using node 6. The rotation DOF in the beam element at node 6 has no corresponding DOF in the 2-D plane elements. We connect the rotation DOF by constraining all the *u* DOF of nodes 3 through 9 as a plane rotating about node 6. This agrees with the assumption for beam theory that plane

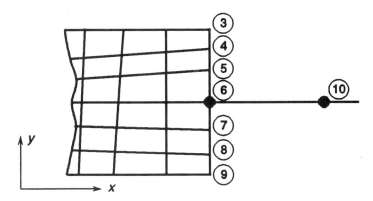

Figure 11-11. Constraint of a Beam Element to 2-D Plane Elements

cross sections remain plane during deformation. There will be six constraint equations to apply. They are

$$u_3 = u_6 - (y_3 - y_6)\theta_6$$

$$u_4 = u_6 - (y_4 - y_6)\theta_6$$

$$u_5 = u_6 - (y_5 - y_6)\theta_6$$

$$u_7 = u_6 + (y_6 - y_7)\theta_6 \qquad (11.1)$$

$$u_8 = u_6 + (y_6 - y_8)\theta_6$$

$$u_9 = u_6 + (y_6 - y_9)\theta_6 .$$

Each of these equations constrains the u displacement component to the value of the θ_6 rotation and the u_6 translation. Effectively, this eliminates the system equations associated with these u DOF. The approximate deformation of the interface is shown in Figure 11-12.

We could attach the beam by only using one of these equations, but physically this would have caused all the bending moment and corresponding rotation of the connection to occur through the single connection. This would not have approximated the uniform beam cross section of the wrench handle very well. A deformed shape plot of such a connection could look like that in Figure 11-13.

The program does not discard the DOF and their equations, but it combines them with other equations to reduce the number of equations. This reduction may be illustrated by a simple example. Take a 3 DOF system given by equation (11.2). Constrain u_3 to be equal to u_2. This results in addition of the terms in the stiffness matrix in column 3 to the terms in column 2. The connection between u_3 and u_2 is now a rigid link

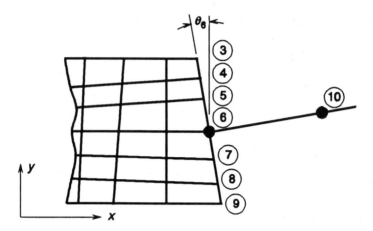

Figure 11-12. Approximate Deformed Shape of Constrained Interface

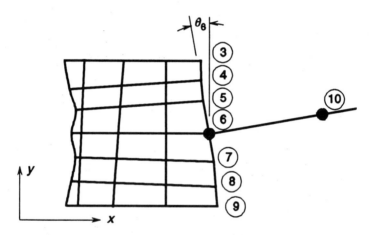

Figure 11-13. Approximate Deformed Shape of One Element Constraint

$$
\begin{bmatrix}
K_{11} & K_{12} & K_{13} \\
K_{21} & K_{22} & K_{23} \\
K_{31} & K_{32} & K_{33}
\end{bmatrix}
\begin{Bmatrix}
u_1 \\
u_2 \\
u_3
\end{Bmatrix}
=
\begin{Bmatrix}
F_1 \\
F_2 \\
F_3
\end{Bmatrix}
\qquad (11.2)
$$

so any external forces on DOF u_3 must transfer to u_2. Accomplish this by addition of equation 3 above to equation 2, and the system of equations reduces to (11.3).

$$
\begin{bmatrix} K_{11} & K_{12} + K_{13} \\ K_{21} + K_{31} & K_{22} + 2K_{23} + K_{33} \end{bmatrix} \begin{Bmatrix} u_1 \\ u_2 \end{Bmatrix} = \begin{Bmatrix} F_1 \\ F_2 + F_3 \end{Bmatrix} \tag{11.3}
$$

The number of equations has reduced by one, and the structure stiffness matrix is still symmetric. Each constraint equation written and input to the model relating two DOF reduces the total DOF and number of system equations by one.

11.4 Subregion Modeling

Many occasions arise in finite element analysis where the solution is good over most of the model but needs more refinement and convergence in one or two local areas. On other occasions the model may be highly complex with many components and refinement in all the components soon becomes impractical. These conditions call for an ability to model a subregion separately that includes the influence from the overall model.

The direct approach to refine and resolve the whole model soon becomes inefficient and may become inaccurate or impractical. Refinement and resolution of the whole model begins to become costly since refinement usually tends to propagate throughout the model. If continued refinement produces large variations in mesh density over the whole model, then numerical ill-conditioning of the equations may begin to lessen the overall accuracy. An initially complex model quickly becomes unmanageable with refinement even though most commercial finite element codes can handle tens of thousands of equations.

There are only a few methods for performing subregion analysis. An approach using constraint equations for local refinement within a subregion without propagation throughout the whole model may be taken [11.8, 11.9]. Another approach is by use of the displacement solution from the whole model along a specified boundary of the subregion in a separate subregion model [11.10, 11.7]. A zooming method identifies the subregion boundary then eliminates all the internal DOF in the whole model outside the subregion and then refines in the subregion [11.11]. The specified boundary stiffness/force method extracts the stiffness coefficients and the internal nodal forces from the whole model solution acting on the subregion boundary [11.12]. These become the boundary conditions on a separate subregion model.

As an example, consider the stress concentration problem of a hole in the center of a thin strip under tension load with the quarter section model shown in Figure 11-14. This is the case study model from Chapter 5. In

Chapter 5 we used the direct approach to refine and converge to an accurate solution. That is the best approach to this problem because it has a simple geometry and converges with a few steps of refinement. We use the case again here simply to illustrate the subregion modeling approaches.

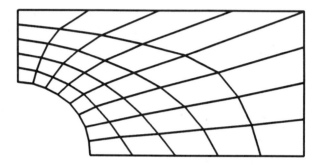

Figure 11-14. Thin Strip With a Center Hole Under Tension Load

The obvious refinement needed for this solution is at the top of the vertical hole diameter. Application of the linear constraint (LC) equation method would begin by selection of the subregion boundary that contains the area of mesh refinement. In this example the element boundaries identified with heavy lines shown in Figure 11-15 could be the subregion boundary.

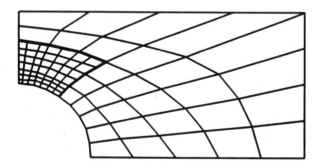

Figure 11-15. Subregion Refinement Using Constraint Equations

The refined mesh in the subregion is also shown. Since multiple nodes along an element edge do not form a compatible connection of linear elements, we define constraint equations for the extra displacement DOF of the added nodes. The displacements of the new nodes along the old element edges are set equal to linearly interpolated values between the values for the corner nodes of the old elements. This model is then run and checked for convergence to decide if we need additional refinement. This method works well if the boundary is sufficiently far away from the area of rapid stress gradients. However, it is not a great deal more efficient than the direct refinement approach.

The specified boundary displacement (SBD) method also may use a subregion boundary similar to that above. In this method the subregion becomes a separate model as illustrated in Figure 11-16.

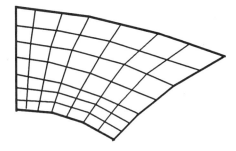

Figure 11-16. Subregion Model For The SBD Method

In this model all nodes on the subregion boundary interface with the whole model have their displacement specified from the first whole model solution. These new nodes do not have to be at any of the whole model nodes since we may interpolate the displacements from the element interpolation functions for any coordinate position within the model.

The biggest advantage of this method is that the system equations only involve the subregion model once the initial whole model solution is done. This method is sensitive to the location of the boundary around the subregion. If it is too close to the area of high gradients, large errors may result from relatively small errors in the boundary displacements. It is also necessary to apply values of boundary displacements with the highest precision available in the computer word. Rounding off the displacement values from the 13-14 significant digits in machine double precision to 5-6 digits in normal output listings can completely invalidate the results.

Application of the zooming (ZM) method using the subregion boundary of Figure 11-15 yields the model shown in Figure 11-17. Condensation of the internal nodes outside the subregion boundary reduces the number of system equations. However, the computing effort to perform the condensation is significant. This method will pay off if we perform several steps of refinement in the subregion as opposed to the direct refinement approach.

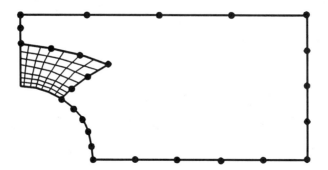

Figure 11-17. Subregion Refinement Using Zooming

The specified boundary stiffness/force (SBSF) method also uses a separate subregion model like the one in Figure 11-16. We extract from the whole model the stiffness coefficients at the interface nodes due to the elements outside the subregion and apply them as boundary stiffnesses on the subregion model. The internal node forces acting on the subregion boundary from the elements outside the boundary become applied loads to the subregion model. We constrain any new nodes added to the interface for refinement to the displacements of the old nodes on the interface, since there is no way to interpolate stiffness coefficients along the old element edges.

The SBSF method shares the advantage of the SBD method in only having system equations involving the subregion model without the computation effort required for condensation of external DOF in the ZM method. The SBSF method is much less sensitive to the location of the boundary around the subregion than the SBD method.

Each of these methods requires that the computer code be able to perform the necessary operations for the method to execute successfully. Local refinement using the LC method requires that the code have ability to input constraint equations. The SBD method can be done manually if the displacement output is available in full machine double precision and input to the separate subregion model with the same precision. It is better if the code has an automated implementation to interpolate for displacements that can be applied to the separate model at full precision. The ANSYS [11.13] code has this implementation. Codes which have a substructuring capability that condenses all internal DOF for a defined substructure can perform zooming subregion analysis. The SBSF method requires access to the structure stiffness matrix coefficients and internal node forces, and that is usually not available in most codes. This method is a recent development not yet implemented into any commercial codes.

11.5   Adaptive Meshing

Adaptive meshing is an effort toward automatic convergence of finite element analyses through mesh refinement or remeshing a model based on error estimates. There is very active research going on in this area [11.14]. The two key components here are a valid error estimate [11.15] for the current solution and availability of an automatic mesh generator [11.16]. Based on the error estimate and automatic mesh generation, the analysis can continue to rerun until it reaches a satisfactorily low error. This is obviously very computationally intensive. However, given the continued increase in computer performance, development of reliable error estimates, and development of automatic mesh generators with no user interaction required, adaptive meshing will become a common method for producing accurate analyses.

Valid and practical error estimates are critical for reaching specified standards of accuracy with finite element codes for practical engineering analysis. The error estimates may express the global error of strain energy density for the whole structure, as well as local estimates at the element level. Error expressions for other variables such as displacement or stress are also useful. Many of the mathematical error prediction techniques are already available and expressed in engineering formulations for use. These are typically complex and costly to perform, so the search continues for more practical measures. The computational cost is a primary factor slowing the inclusion of error estimating in commercial codes. However, a few codes have done so in recent versions, and it will become a common feature soon.

The goal of adaptive meshing is to reduce the discretization error by increasing the number of DOF in areas where the initial mesh has excessive error. Three forms of adaptive refinement exist. The most common and direct is *h-refinement*, which reduces element sizes in the identified areas. In *p-refinement* [11.17] the polynomial order of approximation increases in elements in identified areas. A combination of h- and p-refinement helps to speed up convergence. Use of h-refinement calls for an automatic mesh generator that can use the input of local error level throughout the mesh to regenerate a new mesh without user intervention. We need hierarchic element formulations up to about tenth-order polynomials within the finite element code to use p-refinement.

One of the first commercial programs to incorporate an adaptive meshing strategy is the I-DEAS code [11.18]. In its initial implementation, this code did not use an error estimate for the basis of refinement. It tried to level the strain energy content of all elements throughout the mesh. The intent here is to reduce the error by making the error more uniform over the entire mesh. Elements having high values of strain energy will either subdivide or the program will move their nodes to reduce the element size thereby reducing their strain energy. The program automatically does the remeshing, but the user exercises control over the process by performing the analysis then executing the adaptive remeshing step. There are several parameters which are also under user control such as the percentage of elements the program may subdivide, whether the program may move the nodes or not, and other factors of the algorithm. An example of this type of adaptive meshing follows.

The case study is a tensile bar with a center hole that we studied in Chapter 5, where convergence came by rebuilding the model with refinement where the stress gradient was highest. The free-meshing option of I-DEAS created the starting model for this problem with a global element size of 0.5 inch. The quarter section model in Figure 11-18 has symmetry boundary conditions on the left edge and bottom edge of the model. A uniform tensile stress acts on the right edge.

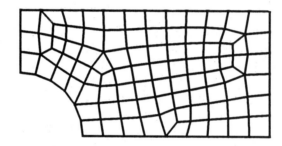

Figure 11-18. Starting Model for Adaptive Meshing Case Study

This mesh is visually appealing with a few distorted elements, but this illustrates the mesh generation flexibility provided with the free-meshing algorithm. Only the boundary geometry and a global element size were input to perform the mesh generation. Specifying local element sizes could provide refinement in the expected high stress area, but the aim here is to show the adaptive meshing procedure and see how quickly it reaches convergence.

The first analysis produced a reasonable deformed shape shown in Figure 11-19, and the x stress distribution in Figure 11-20. The maximum stress reported is 20.4 kpsi and with an average stress on the net cross section of 10 kpsi yields a stress concentration factor of 2.04. The theoretical stress concentration factor is 2.16 so there is a 6 percent error.

Since this adaptive meshing procedure tries to level the strain energy per element, the contour plot of strain energy per element is given in Figure 11-21. Notice that the range of element strain energy is from 1.2 to 75.9 10^{-2} in-lb. Element subdivision or reduction through node movement

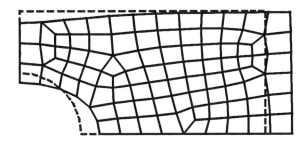

Figure 11-19. Deformed Shape Plot of the Starting Model

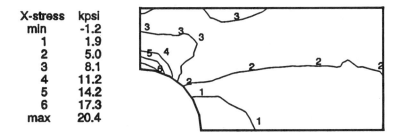

X-stress	kpsi
min	-1.2
1	1.9
2	5.0
3	8.1
4	11.2
5	14.2
6	17.3
max	20.4

Figure 11-20. Contour Plot of σ_x in the Tensile Bar

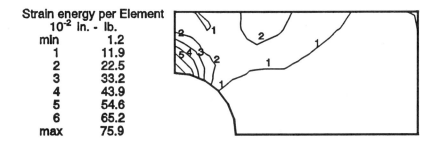

Strain energy per Element 10^{-2} in. - lb.	
min	1.2
1	11.9
2	22.5
3	33.2
4	43.9
5	54.6
6	65.2
max	75.9

Figure 11-21. Contour Plot of Strain Energy per Element

will occur inside the regions of the higher contours. Execution of the adaptive meshing stage produced the mesh in Figure 11-22. The changed elements are mostly inside the contour level 1 along the element edges with some node movement as well.

Analysis of this mesh gave improved results for the stress and a reduction in the range of strain energy per element. The x stress distribution is shown in Figure 11-23. The maximum value is 21.7 kpsi giving a stress concentration factor of 2.17, and the error reduces to 0.5 percent.

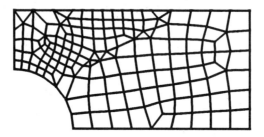

Figure 11-22. Mesh 2 of the Tensile Bar

X - stress	kpsi
min	-1.2
1	2.1
2	5.4
3	8.6
4	11.9
5	15.2
6	18.4
max	21.7

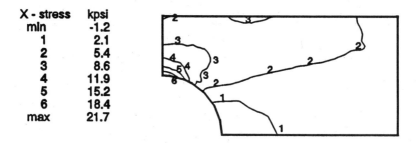

Figure 11-23. Mesh 2 Contour Plot of σ_x in the Tensile Bar

The strain energy plot for mesh 2 is in Figure 11-24. The range reduced by a factor of 3, and the area of further refinement is inside the higher contours.

Another execution of the adaptive meshing stage produced the mesh in Figure 11-25 where the areas of refinement are inside the contour level 2. The x stress for this mesh is shown in Figure 11-26 with a maximum of 22.0 kpsi and stress concentration factor of 2.20. The error increased slightly to about 2 percent.

This clearly shows the ability to converge toward the correct solution with adaptive meshing. Even though this procedure did not use error estimates as its basis and the error went up in the third step, it shows the direction in which adaptive meshing is going. It will become a very automated feature of future finite element analyses.

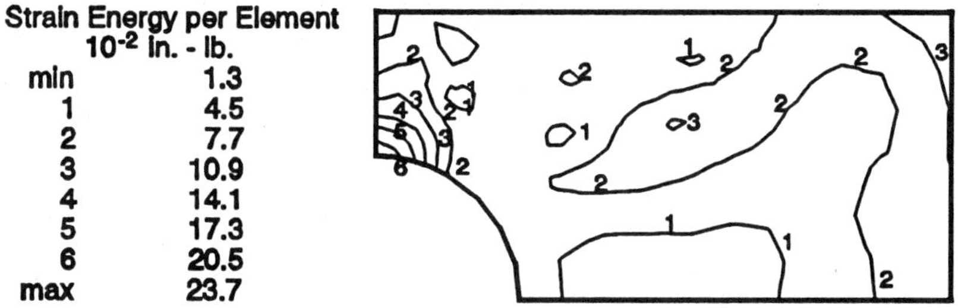

Strain Energy per Element 10^{-2} in. - lb.	
min	1.3
1	4.5
2	7.7
3	10.9
4	14.1
5	17.3
6	20.5
max	23.7

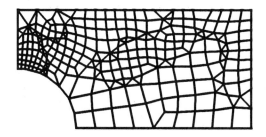

Figure 11-24. Mesh 2 Contour Plot of Strain Energy per Element

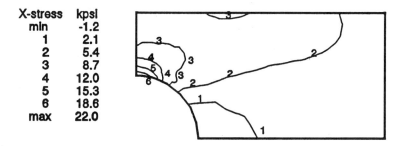

Figure 11-25. Mesh 3 of the Tensile Bar

X-stress kpsi	
min	-1.2
1	2.1
2	5.4
3	8.7
4	12.0
5	15.3
6	18.6
max	22.0

Figure 11-26. Mesh 3 Contour Plot of σ_x in the Tensile Bar

Problems

11.1 The sketch in Figure P11-1 can represent a wheel on a rail or a sphere on a plane. Examine the contact stress problem in both cases by the finite element method using 2-D plane elements for the wheel-rail problem and axisymmetric elements for the sphere-plane problem. The problem is nonlinear in that the area of contact is a function of the load level, so select a load that will produce a contact area several times larger than the local element size at the contact interface. You may find a good coverage of the Hertzian contact stress problem in [11.19] or other mechanics of materials or machine design books. Solution of this problem is easier with a computer program with interface or gap elements, but it may be done by use of very short truss elements with manual adjustment of model conditions for each load application. You may choose any material and specific sizes of interest.

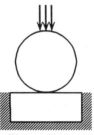

Figure P11 - 1.

11.2 Repeat Problem 7.10 (page 179) using gap or interface elements along the gasket boundaries, and increase the internal pressure until partial separation of the gasket-head interface occurs.

11.3 Repeat Problem 5.5 shown in Figure P11-3 using a combination of beam elements for the straight segments and 2-D plane elements in the stress concentration zones.

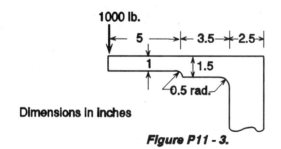

Figure P11 - 3.

11.4 Repeat Problem 5.15 shown in Figure P11-4 using a combination of beam elements for the straight segments and 2-D plane elements in the stress concentration zones.

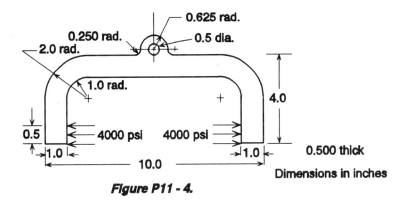

Figure P11 - 4.

11.5 Repeat Problem 6.2 shown in Figure P11-5 using a combination of plate or shell and 3-D solid elements. The solid elements should model the transition corner. Compare with the solution to problem 6.2 if you worked it, especially with respect to the torsional response of the extended arm.

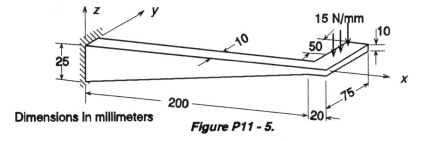

Figure P11 - 5.

11.6 Use some subregion modeling procedure to evaluate the stresses locally around the highest loaded bolt or highest stress area of Problem 5.4 shown in Figure P11-6.

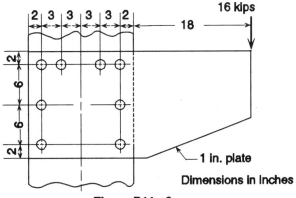

Figure P11 - 6.

11.7   Use some subregion modeling procedure to evaluate the stresses in the stress concentration areas of Problem 5.5 shown in Figure P11-7.

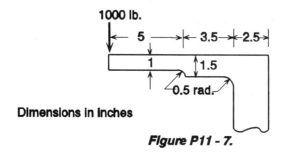

Dimensions in inches

Figure P11 - 7.

11.8   Use an adaptive meshing procedure to solve the same tensile bar with central hole problem illustrated in Figures 11-18. through 11-26. The geometry and symmetric model section appear in Figure P11-8.

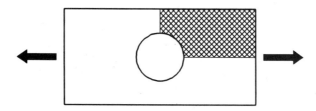

Figure P11 - 8.

References

11.1   Cook, R. D., Malkus, D. S., and Plesha, M. E., *Concepts and Applications of Finite Element Analysis*, Third Edition, John Wiley and Sons, New York, 1989.

11.2   Hallquist, J. O., "A Numerical Treatment of Sliding Interfaces and Impact," in *Computational Techniques for Interface Problems*, K. C. Park and D. K. Gartling (eds.), AMD vol. 30, ASME, New York, 1978.

11.3   Bathe, K. J., and Chaudhary, A., "A Solution Method for Planar and Axisymmetric Contact Problems," Int. J. Numerical Methods in Engineering, Vol. 21, pp. 65-88, 1985.

11.4   Rogers, C. R., *ANSYS Engineering Analysis System Techniques*, Swanson Analysis Systems, Inc., Houston, PA, 1983.

11.5 Wilson, E. A. and Parsons, B., "Finite Element Analysis of Elastic Contact Problems Using Differential Displacements," Int. J. Numerical Methods in Engineering, Vol. 2, pp. 387-395, 1970.

11.6 Stadter, J. T. and Weiss, R. O., "Analysis of Contact Through Finite Element Gaps," Computers and Structures, Vol. 10, pp. 867-873, 1979.

11.7 Kelley, F. S., "Simplified Finite Element Analysis of Stress Discontinuities in Shell Structures," Paper No. 84-PVP-86, ASME, New York.

11.8 Carey, G. F., "A Mesh-Refinement Scheme for Finite Element Computations," Computer Methods in Applied Mechanics and Engineering, Vol. 7, pp. 93-105, 1976.

11.9 Schwartz, D. J., "Practical Analysis of Stress Raisers in Solid Structures," 4th Int. Conf. on Vehicle Structural Mechanics, SAE, Warrendale, PA, Nov. 1981.

11.10 Kelley, F. S., "Mesh Requirements for the Analysis of a Stress Concentration by the Specified Boundary Displacement Method," Proc. 2nd International Computers in Engineering Conference, ASME, Aug. 1982.

11.11 Hirai, I., Wang, B. P. and Pilkey, W. D., "An Efficient Zooming Method for Finite Element Analysis," Int. J. Numerical Methods in Engineering, Vol. 20, pp. 1671-1683, 1984.

11.12 Jara-Almonte, C. C. and Knight, C. E., The Specified Boundary Stiffness/Force, SBSF, Method for Finite Element Subregion Analysis," Int. J. Numerical Methods in Engineering, Vol. 26, pp. 1567-1578, 1988.

11.13 *ANSYS Engineering Analysis System*, Swanson Analysis Systems, Inc., Houston, PA.

11.14 Babuska, I., Zienkiewicz, O. C., Gago, J. and Oliveira, E. R., (eds) *Accuracy Estimates and Adaptive Refinement in Finite Element Computations*, John Wiley and Sons, New York, 1986.

11.15 Zienkiewicz, O. C. and Zhu, J. Z., "A Simple Error Estimator and Adaptive Procedure for Practical Engineering Analysis," Int. J. Numerical Methods in Engineering, Vol. 24, pp. 337-357, 1987.

11.16 Zhu, J. Z., and Zienkiewicz, O. C., "Adaptive Techniques in the Finite Element Method," Communications in Applied Numerical Methods, Vol. 4, pp. 197-204, 1988.

11.17 Szabo, B. and Babuska, I., *Finite Element Analysis*, John Wiley and Sons, New York, 1991.

11.18 *I-DEAS, Integrated Design Engineering and Analysis System*, Structural Dynamics Research Corporation, Cincinnati, Ohio.

11.19 Boresi, A. P., and Sidebottom, O. M., *Advanced Mechanics of Materials*, 4th Edition, John Wiley and Sons, New York, 1985

C H A P T E R   12

HEAT TRANSFER AND THERMAL STRESS

The temperature distribution within a mechanical component may play an important role in its ability to perform without failure. Obviously the material strength and other properties may be a function of temperature especially at high temperatures. However, even at temperatures below property degradation levels, a variable temperature distribution can produce thermal stress. Also, a restraint of the component during heating can produce thermal stress. Thermal stresses may combine directly with stresses from mechanical loading by superposition in linear analyses. The total stress then determines the potential for success in the component design.

The first step in finding the thermal stress is to determine the temperature distribution. Once the temperature distribution is known computation of the thermal stress then follows directly. This chapter will discuss the application of the finite element method for conduction heat transfer. Heat transfer may be done by other numerical methods, but to use the temperature distribution to find thermal stress it is better if done with finite elements. In fact, usually we use the same element mesh to model both the heat transfer and stress analysis problem. Sometimes though a one-dimensional heat transfer solution will produce two- or three-dimensional stress states. In these cases, we may use separate models and map the temperature distribution over the area or volume of the solid mesh for stress analysis.

12.1   The Governing Equations

The basic theory governing heat conduction in solids is the Fourier law of heat conduction. The law states that the amount of heat conducted between two points is proportional to the area of conduction, the temperature difference, and the time, and it is inversely proportional to the distance between the two points. Heat flux is a measure of the heat flow intensity. The definition of heat flux is the amount of heat flowing per unit area and unit time.

Adopting the notation of Incropera and DeWitt[12.1], we write the Fourier law of heat conduction as

$$q'' = -k\nabla T \qquad (12.1)$$

where, q'' is the heat flux vector, k is the thermal conductivity, and ∇T is the vector gradient of the temperature field. The heat flux and temperature may both be functions of space and time. In steady-state conduction with no heat sinks or sources within the body, the heat flux must be a constant to satisfy the conservation of energy. Since it is constant, the spatial derivative or gradient of the heat flux above is zero which provides the governing differential equation for three-dimensional, steady-state conduction.

$$\frac{\partial}{\partial x}\left(k\frac{\partial T}{\partial x}\right) + \frac{\partial}{\partial y}\left(k\frac{\partial T}{\partial y}\right) + \frac{\partial}{\partial z}\left(k\frac{\partial T}{\partial z}\right) = 0 \qquad (12.2)$$

Heat sources, or sinks within the body and the heat stored in the transient by the material's heat capacity cause changes to the heat flux through the body . We may simply add these terms to the equation to satisfy the conservation of energy, and if the conductivity is constant throughout then the equation becomes

$$k\frac{\partial^2 T}{\partial x^2} + k\frac{\partial^2 T}{\partial y^2} + k\frac{\partial^2 T}{\partial z^2} + \dot{q} = \rho c_p \frac{\partial T}{\partial t} \qquad (12.3)$$

where, $\dot{q}$ is the heat rate per unit volume generated by the source or absorbed by the sink, ρ is the material mass density, c_p is the material specific heat, and t is the time variable.

The equation above is the general form of the heat diffusion equation. Most of the discussion in this chapter will focus on finding the solution in steady-state conditions by setting the right hand side of the equation to zero. We may obviously simplify the equation further to two-dimensional or one-dimensional forms.

We can solve the equation only by application of the proper boundary conditions to the equation. These are physical conditions existing on the boundary of the region. Three types of boundary conditions are common in heat transfer. They are specified temperature, specified constant heat flux, and convection. One of these conditions may be active for any portion of the boundary, but the complete boundary of the region must have specified boundary conditions.

Boundary conditions are shown in Figure 12-1, where boundary segment S_1 has a fixed temperature, segment S_2 has a fixed heat flux, and segment S_3 has a convection condition. Portions with a specified temperature simply assign the solution values on those portions since the solution variable is temperature. Portions with specified heat flux establish the gradient of the temperature distribution normal to the boundary surface. An insulated boundary has a heat flux of zero, and therefore the temperature gradient normal to the boundary is zero. Portions with convection heating or cooling are, in effect, a specified heat flux with the flux value being a function of the surface temperature and fluid temperature. Nonlinear thermal radiation boundary conditions can be approximated linearly using the concept of a radiation heat transfer coefficient [12.1].

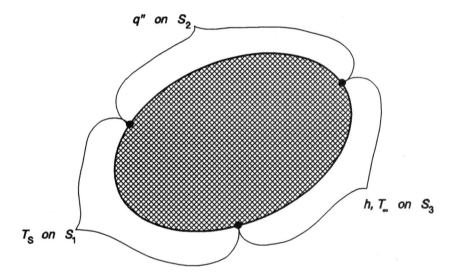

Figure 12-1. Boundary Conditions For Heat Conduction in the Domain

Once we get the temperature distribution, the thermal stress calculation follows directly. Thermal stresses due to temperature changes build up whenever there is a nonlinear temperature gradient, a displacement restraint, or the material is anisotropic.

Thermal effects are accounted for in the stress-strain relations by including thermal expansion as an initial strain. The relations become

$$\{\sigma\} = [E](\{\epsilon\} - \{\epsilon_0\}) \tag{12.4}$$

where, $\{\sigma\}$ are the stress components, $[E]$ is the material stiffness matrix, $\{\epsilon\}$ are the mechanical strain components, and $\{\epsilon_0\}$ are the initial or thermal strains. The initial strains due to thermal expansion have only normal strain components. The values of 3-D components become $\epsilon_x = \epsilon_y = \epsilon_z = \alpha T$ and $\gamma_{xy} = \gamma_{xz} = \gamma_{yz} = 0$, where α is the thermal expansion coefficient, and T is the temperature change relative to the reference unstressed temperature.

The usual approach is to calculate the stresses that would exist if the element could not expand with the thermal strain. We then calculate the node point loads caused by these stresses and apply them as external loads. The model responds to these loads as a normal static load case. After solving for the displacements, we then use the stress-strain relations above to calculate the stress in each element.

If the model is uniformly heated so that T is spatially constant and the material is isotropic, then the stress will be zero throughout if there is no restraint of displacements. This is because the displacements of the nodes due to the thermal "loads" yield mechanical strains that exactly equal the thermal strains for a net strain of zero. Under the same conditions if the displacements are completely restrained, then the mechanical strains will be zero and a state of compressive stress equal to the $[E]\{\epsilon_0\}$ product exists throughout. We also note that if T is spatially linear with no displacement restraints the stress is also zero [12.2].

We can illustrate the approach through a one-dimensional example. Taking the two-spring model of Chapter 1 to model a rod, we remove the applied load, F, and heat each element to temperature T. Using an area, A, an elastic modulus, E, and element length, $L/2$, the stiffness is $k = 2AE/L$. The initial strain, $\epsilon_0 = \alpha T$ would give a stress of $E\alpha T$ if the element is fully restrained. The resulting element nodal loads would be $-E\alpha TA$ on the left node and $+E\alpha TA$ on the right node. The system equations corresponding to equation (1.10) when the left node of the structure model is fixed and the right node is free become

$$\begin{bmatrix} 2k & -k \\ -k & k \end{bmatrix} \begin{Bmatrix} u_2 \\ u_3 \end{Bmatrix} = \begin{Bmatrix} 0 \\ E\alpha TA \end{Bmatrix}. \tag{12.5}$$

Solving the equations gives

$$u_2 = \alpha TL/2 \quad \text{and} \quad u_3 = \alpha TL .\qquad(12.6)$$

Calculating the stresses from equation (12.4) gives

$$\sigma_1 = E\left(\frac{u_2 - 0}{L/2} - \alpha T\right) = 0$$

$$(12.7)$$

$$\sigma_2 = E\left(\frac{u_3 - u_2}{L/2} - \alpha T\right) = 0 .$$

It is easy to see that if node 3 is restrained then the displacement of node 2 is zero and the thermal stress becomes $\sigma = E\alpha T$ in each element.

12.2   Element Formulation

We may form heat conduction elements for problems in one-dimension, two-dimensions, axisymmetric, or three-dimensions as was done with structural elements. These formulations are then available in the element library of a finite element code for selection by the user. We will form the one-dimensional element directly, as done by Huebner and Thornton [12.3], to help understand the physical concepts of the problem without the mathematical complexities.

Consider one-dimensional steady state heat flow through a layered plate or wall as sketched in Figure 12-2. Each layer has a different value of

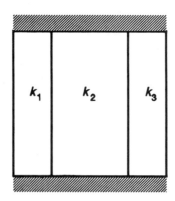

Figure 12-2. Heat Flow Through a Layered Plate or Wall

thermal conductivity, and three heat elements in series will suffice for the model as in Figure 12-3.

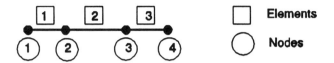

Figure 12-3. Three Element Finite Element Model

To create the finite element model, we need an element formulation. An element having a conductivity, k_p, with two nodes is shown in Figure 12-4, where p may denote any element in the model. We assume the element has a constant cross section area.

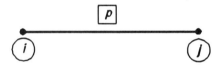

Figure 12-4. A Two-Node One-Dimensional Element

Applying the Fourier law of heat conduction yields equation (12.8), where, x is the local coordinate along the length of the element.

$$q_x'' = -k_p \frac{dT}{dx} \tag{12.8}$$

Separating the variables and integrating gives

$$q_x'' \int_{x_i}^{x_j} dx = -k_p \int_{T_i}^{T_j} dT . \tag{12.9}$$

The result is equation (12.10), where subscripts i and j denote nodal values, and l_p is the element length.

$$q_x'' l_p = -k_p(T_j - T_i) \tag{12.10}$$

The heat flux is positive when flowing into the element through the node. The heat flux at node i is in equation (12.11), and the flux at node j being the negative of q_{ip}'' is in equation (12.12).

$$q_{ip}'' = -\frac{k_p}{l_p}(T_j - T_i) \tag{12.11}$$

$$q_{jp}'' = \frac{k_p}{l_p}(T_j - T_i) \tag{12.12}$$

The cross section area of the element is constant, but to allow for the area to change from one element to the next we reform the equations to relate total heat flow rate through the element. Multiplying both sides by the element area gives

$$q_{ip} = q_{ip}'' A_p = -\frac{A_p k_p}{l_p}(T_j - T_i)$$

$$q_{jp} = q_{jp}'' A_p = \frac{A_p k_p}{l_p}(T_j - T_i) \tag{12.13}$$

where, q_{ip} and q_{jp} are the heat flow rates at nodes i and j respectively, and A_p is the element cross section area. Defining the term

$$k_p = \frac{A_p k_p}{l_p} \tag{12.14}$$

the element equations in matrix form become

$$\begin{bmatrix} k_p & -k_p \\ -k_p & k_p \end{bmatrix} \begin{Bmatrix} T_i \\ T_j \end{Bmatrix} = \begin{Bmatrix} q_{ip} \\ q_{jp} \end{Bmatrix} . \tag{12.15}$$

We can assemble the system equations for steady-state heat flow by requiring that the sum of the internal heat rates at a node equal any applied external nodal heat rate. The requirement is written in equation (12.16), where, q_{ie} is the external nodal heat rate and m is the number of elements connected to node i.

$$\sum_{p=1}^{m} q_{ip} = q_{ie} \tag{12.16}$$

Now turning to the finite element model in Figure 12-3, the element equations become for element 1,

$$\begin{bmatrix} k_1 & -k_1 \\ -k_1 & k_1 \end{bmatrix} \begin{Bmatrix} T_1 \\ T_2 \end{Bmatrix} = \begin{Bmatrix} q_{11} \\ q_{21} \end{Bmatrix} \tag{12.17}$$

for element 2,

$$\begin{bmatrix} k_2 & -k_2 \\ -k_2 & k_2 \end{bmatrix} \begin{Bmatrix} T_2 \\ T_3 \end{Bmatrix} = \begin{Bmatrix} q_{22} \\ q_{32} \end{Bmatrix} \tag{12.18}$$

and for element 3,

$$\begin{bmatrix} k_3 & -k_3 \\ -k_3 & k_3 \end{bmatrix} \begin{Bmatrix} T_3 \\ T_4 \end{Bmatrix} = \begin{Bmatrix} q_{33} \\ q_{43} \end{Bmatrix} . \tag{12.19}$$

Now summing the heat rates at each node yields

NODE 1 $\Rightarrow$ $k_1 T_1 - k_1 T_2 = q_{11} = q_{1e}$

NODE 2 $\Rightarrow$ $-k_1 T_1 + k_1 T_2 + k_2 T_2 - k_2 T_3 = q_{21} + q_{22} = q_{2e}$

NODE 3 $\Rightarrow$ $-k_2 T_2 + k_2 T_3 + k_3 T_3 - k_3 T_4 = q_{32} + q_{33} = q_{3e}$ $\tag{12.20}$

NODE 4 $\Rightarrow$ $-k_3 T_3 - k_3 T_4 = q_{43} = q_{4e}$.

Arranging these equations in matrix form yields equation (12.21).

In every problem, we must specify flux or convection boundary conditions to determine the heat rate vector for all nodes except those with a specified temperature. There must be at least one specified temperature

$$
\begin{bmatrix}
k_1 & -k_1 & 0 & 0 \\
-k_1 & (k_1 + k_2) & -k_2 & 0 \\
0 & -k_2 & (k_2 + k_3) & -k_3 \\
0 & 0 & -k_3 & k_3
\end{bmatrix}
\begin{Bmatrix}
T_1 \\
T_2 \\
T_3 \\
T_4
\end{Bmatrix}
=
\begin{Bmatrix}
q_{1e} \\
q_{2e} \\
q_{3e} \\
q_{4e}
\end{Bmatrix}
\qquad (12.21)
$$

or one convection condition to set the reference temperature in a steady-state analysis.

The reason for setting the reference temperature is that the conductivity matrix is singular before application of one temperature boundary condition. Physically, this means in this example, that the temperature profile that fit through the four nodes is not locked to any temperature scale. We can only find three of the values relative to the fourth. If one of the nodes has a heat source which holds it at a specified temperature, it becomes the reference temperature.

Once we have a reference temperature, it may be set to zero by a scalar shift of all the temperatures. A zero temperature value has the effect of cutting the corresponding column in the conductivity matrix, and we may discard the corresponding equation row. This reduces the conductivity matrix in this example to a 3 x 3, or three system equations to solve. Of course, the reference temperature may remain nonzero in which case the product of the conductivity column with the reference temperature moves to the right side of the equation and the row is discarded. This still leaves three equations to solve.

A convection boundary condition also may provide the reference temperature which is the constant fluid temperature of the convecting fluid. The convection condition may be expressed by

$$
q_c = Ah(T_s - T_f) \qquad (12.22)
$$

where, q_c is the convection heat rate, A is the surface area, h is the convection coefficient, T_s is the surface temperature, and T_f is the fluid temperature. The fluid temperature remains constant while the surface temperature is a variable. To include the convection in the system equations, the term AhT_s must move to the left side of the equations and Ah adds into the conductivity matrix at the location of the surface node. The other term, AhT_f, remains in the heat rate column vector on the right side of the system equations.

With input of specific property and geometric values, the equations are easy to solve. In this one-dimensional element, convection may occur from the lateral surface as well as the element ends at the nodes. In that case the heat rate from the lateral surface must be split between the two nodes of the element based on the temperature interpolation along the element length.

The other boundary condition that may determine the external nodal heat rate is a specified heat flux. Of course, an insulated boundary corresponds to a zero external heat flux and zero nodal heat rate. Non-zero heat flux may occur on the element ends or on the lateral surface. On the ends, calculate the value by multiplication with the cross section area. On the lateral surface, calculate the value by multiplication with the lateral surface area, and split it between the two nodes of the element as we did for the lateral convection.

These equations and boundary condition application illustrate the general approach. The system equations provide for the solution of node point temperatures using the conductivity matrix augmented by any convection conditions and the boundary conditions converted to node point external heat rates. For two- and three-dimensional solids, elements must account for the spatial variation of temperature within the element domain and provide the methods for computing nodal heat rates from inputs of convection or heat flux.

Two- and three-dimensional elements are not easy to formulate in a direct manner as in the one-dimensional element. Their formulation requires more advanced mathematical concepts. Most commonly we employ calculus of variations to develop the general relations for element formulation. The method of weighted residuals, specifically the Galerkin method, works when we cannot develop a formulation from calculus of variations. For more in-depth study consult Huebner and Thornton [12.3].

This text presents the types of elements available for use, but not their general formulations. For two-dimensional analysis, the triangular element was the first to develop as it was in stress analysis. The triangular element illustrated in Figure 12-5 defines an area bounded by the three sides connecting three node points. Within the element area the temperature assumes a linear distribution given by

$$T = a_1 + a_2 x + a_3 y \tag{12.23}$$

where, T is the temperature of a material point within the element field, x and y are coordinates of the point, and a_i, $i = 1,2,3$ are constant coefficients to determine. The linear function has three undetermined coefficients, and since there are three nodes we relate the coefficients to the node point temperature values.

This triangular element works very well for heat transfer solutions. However, since the triangular element does not work well in stress analysis, a quadrilateral heat transfer element is more appropriate

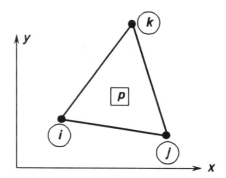

Figure 12-5. Triangular Two-Dimensional Finite Element

whenever we compute thermal stresses from the heat transfer analysis.

The isoparametric formulation presented for stress analysis provides the interpolation formulas needed for the conduction heat transfer element. The quadrilateral element is drawn in Figure 12-6. The assumed temperature field now becomes

$$T = a_1 + a_2\xi + a_3\eta + a_4\xi\eta . \qquad (12.24)$$

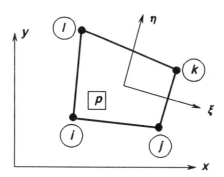

Figure 12-6. Two-Dimensional Quadrilateral Element

This interpolation provides some improvement in approximating the actual temperature distribution in the quadrilateral area over the approximation in the triangle element. More importantly it allows for use of the same element mesh for the conduction heat transfer analysis and the thermal stress analysis that needs the node point temperature distribution as input.

Both the triangle and quadrilateral element types may easily convert to an axisymmetric solid element formulation. Three-dimensional heat transfer element formulation follows a path similar to the stress analysis elements described previously.

Cook [12.4] suggests that the temperature and strain distributions should have close to the same order of approximation to avoid stress errors. As we saw in the previous section, the temperature change throughout the element represents an initial strain. In these corner-noded elements the temperature change or initial strain has a linear or nearly linear distribution. Yet, in the same elements the mechanical strain has a constant or nearly constant distribution. This difference in approximation order between initial strain and mechanical strain can lead to calculation of spurious stresses.

The stress calculation is usually good at the element centroid, but may be in significant error at other locations such as the node points. One approach to alleviate this problem is to use linear, corner-noded elements for the temperature solution and parabolic, midside-noded elements for the stress solution. Fortunately, these errors usually decrease with increased mesh refinement as well.

Wilson and Ibrahimbegovic [12.5] give a new method for stress recovery. It uses a least-square approximation to fit a linear stress distribution which is in microscopic equilibrium and in global equilibrium with the nodal loads. Their examples show excellent results in thermal stress cases. This is a recent development (1990 publication) and may take some time to be widely incorporated into commercial finite element codes.

12.3   The Finite Element Model

In the line element, one-dimensional models there is no reason for element subdivision other than to define the geometry of the structure or to account for lateral convection or heat flux. This is because the solution is exact for the one-dimensional heat flow. However, in the two- and three-dimensional cases we must do successive mesh refinement to achieve an accurate solution. A recommended procedure is to begin the model with a linear triangle or quadrilateral element and refine by subdivision until you get a reasonable solution. You may follow this by changing the element type to one of quadratic order in the final mesh. If the results do not change dramatically in this last step, then you should have the converged solution.

Begin the mesh plan by laying out the model geometry and recognizing any usable symmetries. The plan is further aided through approximate engineering calculations to develop some insight into the expected variation of temperature throughout the body. From these estimates identify the most critical regions in the body (where high heat fluxes and therefore high temperature gradients occur) and plan for adequate mesh refinement in those regions.

Symmetry in two- or three-dimensional problems requires both geometric and boundary condition symmetry.   Enforce symmetry conditions by imposing insulated boundary conditions on all planes of symmetry. An insulated boundary has a heat flux of zero corresponding to the natural boundary condition existing at all nodes. So the insulated boundary condition needs no additional input from the user. For example, a rectangular tube section in Figure 12-7 has a constant temperature on the inner surface and convection from the outer surface. It has vertical and horizontal planes of symmetry through the centroid. A quarter section model is sufficient as in Figure 12-8.

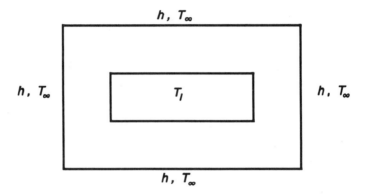

Figure 12-7. Heat Transfer Through a Rectangular Tube

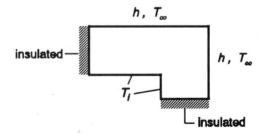

Figure 12-8. One-Quarter Section Model of the Rectangular Tube

The mesh plan for this example should begin with a relatively coarse subdivision with some consideration of the expected temperature variations, and whether we will make a thermal stress calculation. The temperature will have a gradient through the wall thickness and therefore requires mesh subdivision through the wall. There will also be some gradient parallel to the wall but probably not large and therefore requires few element subdivisions except perhaps near the corner. A reasonable starting mesh division based on thermal considerations is shown in Figure 12-9.

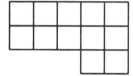

Figure 12-9. Starting Mesh for Heat Transfer in the Section Model

However, if we also want to make a thermal stress calculation, then very serious stress gradients and peak stresses may occur at the inside corner. An inside corner with little or no radius has a very large stress concentration effect. This case requires a more refined mesh near the corner and probably warrants additional refinement in other areas as well. A suggested starting mesh for the heat transfer and thermal stress problem is shown in Figure 12-10.

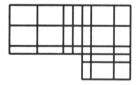

Figure 12-10. Starting Mesh for Thermal Stress in the Section Model

In some cases there may be symmetries in the heat transfer problem that we cannot utilize in the thermal stress analysis. For example, a hexagonal brick as in Figure 12-11 with uniform heat flux passing through one face to the opposite face with all the lateral surface insulated is a one-dimensional heat transfer problem. However, the thermal stress created by this one-dimensional temperature distribution is fully three-dimensional because only the planes passing through the center parallel to the lateral faces and normal to the heat transfer faces are planes of symmetry.

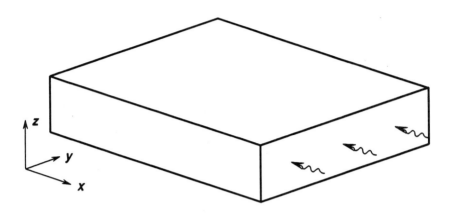

Figure 12-11. A Hexagonal Brick with Uniform Heat Flux

Therefore, we may use a one-dimensional model for heat transfer as in Figure 12-12(a) and a quarter section three-dimensional model for thermal stress as in Figure 12-12(b) or use the same three-dimensional model for both analyses. If we use a one-dimensional heat transfer model, it should match the nodal spacing along the heat flow direction of the three-dimensional stress model. All the nodes in the stress model having the same coordinate location along the heat flow direction as a node in the heat transfer model then have the same temperature. If we use the same 3-D model for both analyses, then we avoid the problem of mapping the temperatures, but it requires considerably more computer time especially in transient cases.

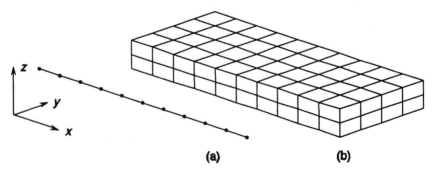

(a)   (b)

Figure 12-12. Finite Element Models of the Brick

12.4   Computer Input Assistance

The preprocessing methods throughout this text for one-, two-, and three-dimensional model development also apply to heat transfer models. The mesh generation of nodes and elements may obviously come from using the same algorithms. Applying boundary conditions is different but we need the same kind of menus and interactivity to make their specification as easy as possible. While in structural problems displacements and forces were the only boundary conditions, there are several different types of initial and boundary conditions required for a conduction heat transfer analysis.

The natural condition of an insulated boundary or a zero heat flux occurs on all model boundaries that have no other specification. Fix all temperatures that stay constant during the analysis by selecting the correct nodes and entering the corresponding temperature. A transient analysis also requires input of the initial temperature distribution which in most cases is a uniform value equal to the reference temperature.

We specify nonzero heat flux conditions by first identifying an area of the surface having a constant heat flux. Identify the area by selecting the element edges along the surface, and then enter the specified heat flux value. The program converts the flux to nodal values of heat rate and adds them to the thermal "load" vector acting on the model.

We specify convection boundary conditions also by first identifying the area making up the convective surface. Then we input the convection coefficient and the ambient fluid temperature. The program converts these to nodal heat rate values as well. The portion of the convection heat rate from the fluid temperature adds into the thermal "load" vector, and the portion from the surface node temperature adds into the conductivity matrix of the model according to equation (12.22).

Programs usually allow specification of a lumped or concentrated heat rate at nodes. Input of lumped values requires the user to make the translation of distributed surface conditions into nodal values. This is similar to the conversion of pressure into concentrated nodal force loading on a structure.

In one- and two-dimensional analyses, heat flux and convection conditions also may come from the other coordinate directions that were excluded in their formulation. A one-dimensional element may have these conditions specified on its surface area lateral to the one-dimensional direction. A two-dimensional element may have these specified normal to its plane surface.

All of the boundary conditions discussed thus far are linear in the steady-state solution or the solution step in a transient analysis. Two nonlinearities which may enter into an analysis are radiation boundary conditions and temperature dependent properties. These require iterative procedures or linearized approximations within a steady-state solution or a solution step in a transient analysis.

12.5   The Analysis Step

The finite element equations for the steady-state problem take the form of equation (12.25) [12.3].

$$[K]\{T\} = \{Q\} \qquad\qquad (12.25)$$

Here, $[K]$, is the conductivity matrix of the model, $\{T\}$, is the node temperature vector, and $\{Q\}$ is the node heat rate or thermal "load" vector.

In the transient case the finite element equations become [12.3]

$$[C]\{\dot{T}\} + [K]\{T\} = \{Q\} \qquad\qquad (12.26)$$

where, $[C]$ is the heat capacitance matrix of the model, and $\{\dot{T}\}$ is the time derivative of the node temperatures. Most programs usually approximate the time derivative by either an explicit or implicit finite difference procedure for marching forward in time.

Solution of the equations result in a set of nodal temperatures in the steady-state case or the consecutive set of nodal temperatures at the chosen time increments for the transient case. Solution of the equations is dependent on proper application of boundary conditions. The steady-state equations require that we specify either the temperature or its derivative with respect to the direction that is normal to the surface over the entire boundary. This translates to all nodes on the boundary surface of any finite element model. The derivative specification is done with either a surface heat flux, surface convection, or nodal heat rate. The transient case, in addition, requires the initial temperature distribution at all nodes in the model, and the boundary conditions may be a function of time.

For each nodal temperature distribution we can determine its corresponding thermal stress field. For steady-state heat transfer this results in only one thermal stress solution. For transient heat transfer each temperature distribution at each time step will have a unique stress distribution. Unless the stress-strain behavior of the material becomes nonlinear, each stress distribution is only dependent on the current temperature distribution. In some problems, the time increment required to find an accurate heat transfer solution may be small and not result in large changes in the stress distribution. In many programs the analyst is free to choose to calculate the thermal stress at longer intervals than at each time increment in the heat transfer solution.

12.6   Output Processing and Evaluation

A program will always create a listing file of all input data and results from the analysis. We should scan this file for any wrong interpretations of data or other input errors. The raw data and results are usually too numerous for evaluating the overall temperature distribution, but we can pick out temperatures at selected locations.

Graphic display of the temperature distributions is usually done with contour plots much like the stress components in a stress analysis. In this case the temperature is a scalar, and so there is only one variable to display. You must examine the distribution and decide if it is reasonable and accurate. Whether it is reasonable or not depends on the problem, so you must be able to judge and see the correlation between the displayed solution and the nature of the problem. We may make some judgment about the accuracy by comparing the mesh density with the distance between contour intervals. Since, a corner-noded element approximates the temperature distribution within the element as nearly linear, then a mesh density which is approximately equal to the contour interval spacing should have good accuracy assuming there are about ten contours covering the full temperature range of the solution.

There are other aspects of the contours to check as well. The contours should become normal to insulated boundaries. They should approach other boundaries with a slope determined by the heat flux at the boundary. Finally, they should be smooth and continuous. Abrupt changes in direction indicate the need for further mesh refinement before accepting the results.

Of course, in a transient solution there may be a display to prepare and evaluate for each time increment or for selected values of time. These plots should show the progression of the temperature distribution as a function of time from the initial conditions. Again you must be able to evaluate the results for reasonableness and accuracy to decide if you need further refinement or if the modeling is correct.

These steps should be enough if only the temperature variation is of interest. However, if we are computing thermal stress, then the mesh may need significant refinement to produce reliable and accurate results. Evaluate the stress results based on the recommendations presented in previous chapters.

12.7   Case Studies

The first case will examine the thermal stresses produced in a thick cylinder under a radial variation of temperature. A long cylinder has a steady-state temperature on the inner wall of 100 °F and an outer wall temperature of 0 °F . We can find the steady-state temperature distribution using an axisymmetric finite element model with a radial subdivision of elements. The basic geometry and contour plot of the temperature distribution is shown in Figure 12-13. The finite element model had 10 elements through the thickness with the top and bottom surfaces insulated because there is no axial heat flow.

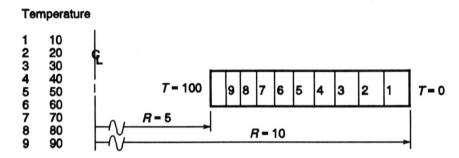

Figure 12-13. Radial Temperature Contour Plot in a Thick Cylinder

In this case since there is only a radial variation of variables, cartesian plots may present clearer graphic display of the data. The plot of temperature versus radial position is given in Figure 12-14.

Actually this heat transfer problem has an exact solution [12.1] which is given by equation (12.27) when the outside temperature is zero, where,

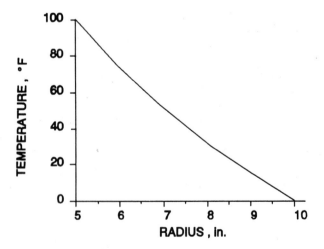

Figure 12-14. Radial Temperature Profile in a Thick Cylinder

$T(r)$ is the temperature at radius r, r_i and r_o are the inside and outside radii, and T_i is the inside wall temperature. The finite element solution above is within 0.15 percent agreement with the theoretical solution.

$$T(r) = \frac{\ln(r/r_o)}{\ln(r_i/r_o)} T_i \qquad (12.27)$$

The finite element model for thermal stress is also one element high by ten elements through the thickness. We fixed all axial (parallel to centerline) nodal displacements to simulate plane strain conditions for a long cylinder. The nodal temperatures from the heat transfer analysis are input to the stress model.

The radial stress through the cylinder wall has the distribution given in Figure 12-15. Note that the radial stress is near zero at the inner and outer wall as it should be. The peak value is 2,160 psi compression just inside the middle of the wall.

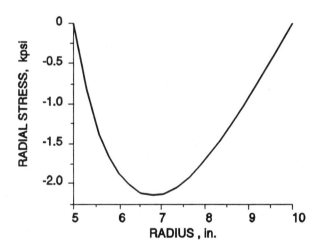

Figure 12-15. Radial Stress Profile in a Thick Cylinder

The axial stress has the distribution given in Figure 12-16. The value ranges from 22,700 psi compression to 3,000 psi tension. This distribution is the result of fixed axial displacement to yield zero axial strain corresponding to plane strain conditions. Of course, there must be some physical restraint to produce this condition. Without axial restraint the axial stress distribution would shift to a self-equilibrating distribution in generalized plane strain conditions. This would reduce the compressive stress and increase the tensile stress to reach axial force equilibrium. In fact, the values at the inside and outside radius must equal the hoop stress values presented below.

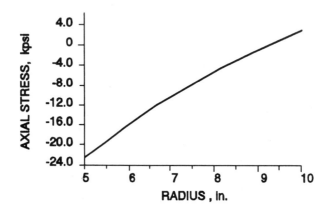

Figure 12-16. Axial Stress Profile in a Thick Cylinder

The hoop stress is shown in Figure 12-17. The value ranges from 15,600 compression to 10,000 psi tension. A theoretical solution for the thermal stresses in this case is also available [12.6]. The results above agree with the theoretical solution within 1.0 percent. This case readily demonstrates that a one-dimensional temperature distribution can produce a three-dimensional state of stress.

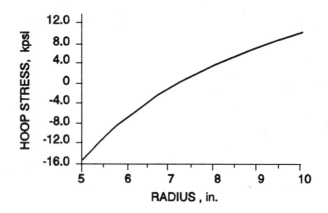

Figure 12-17. Hoop Stress Profile in a Thick Cylinder

The second case also will examine the thermal stresses in a long cylinder. However, this time the cross section will be rectangular and the thermal conditions will be different. The cross section geometry is shown in Figure 12-18. The boundary conditions are uniform convection on the inner and outer surface with a coefficient of 9.0 Btu/(hr·in²·°F) with a fluid temperature of 200 °F on the inside and 0 °F on the outside. These conditions and the geometric symmetry allow the use of a quarter section model corresponding to the hatched area in Figure 12-18.

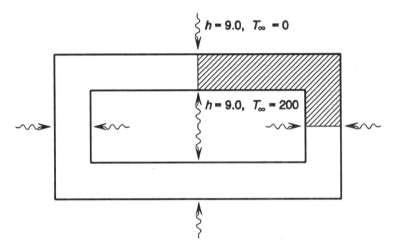

Figure 12-18. Rectangular Cross Section of a Long Cylinder

The first finite element mesh appears in Figure 12-19. There has been no attempt in this mesh to account for the obvious stress concentration at the inside corner of the section. This mesh should serve to identify any

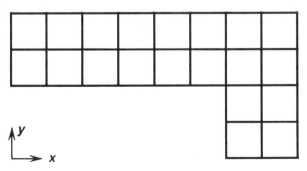

Figure 12-19. Finite Element Mesh 1 of a Rectangular Long Cylinder

local disturbance in the temperature distribution caused by the inside corner. In the heat transfer analysis we applied the inside and outside convection conditions to the respective surfaces and the edges on the planes of symmetry had natural insulated conditions.

A contour plot of the temperature distribution is in Figure 12-20 with the highest temperature on the center of the longest inside wall and the lowest at the outside corner. There is no evidence of any local effects at the inside corner.

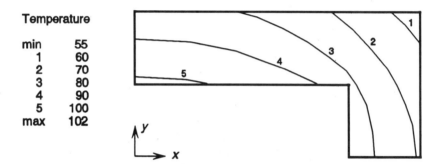

Temperature	
min	55
1	60
2	70
3	80
4	90
5	100
max	102

Figure 12-20. Temperature Distribution From Mesh 1

For the thermal stress analysis we input the nodal temperatures and change the element type to 2-D plane strain using the same mesh. We impose symmetry displacement boundary conditions on the $x = 0$ and $y = 0$ planes, and the reference stress free temperature is taken to be 0 °F. The contour plot of the x stress component is given in Figure 12-21. As expected from the temperature distribution, this stress component exhibits a bending distribution along the long leg of the section. However, this

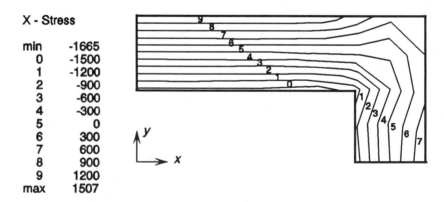

X - Stress	
min	-1665
0	-1500
1	-1200
2	-900
3	-600
4	-300
5	0
6	300
7	600
8	900
9	1200
max	1507

Figure 12-21. The X Stress Component in Mesh 1

mesh is entirely too coarse, since the whole range of tension stress occurs across one element as does the range of compression stress. Also, this model shows almost none of the stress concentration at the inside corner.

We also should note that the x stress should be zero along the outside right edge of the section which is not even closely satisfied in this first model. A similar distribution occurs for the y stress component in the short leg of the section. The Von Mises equivalent stress in this first model is in Figure 12-22. The stress concentration at the inside corner is also missing in this result.

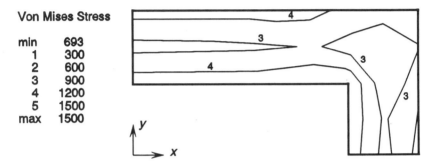

Figure 12-22. The Von Mises Equivalent Stress in Mesh 1

Since the theoretical solution for this problem is unknown, and there are obvious deficiencies in the first mesh solution, it requires further modeling. The second mesh is shown in Figure 12-23. The element size is biased toward smaller elements at the inside corner.

The temperature distribution contour plot for Mesh 2 was virtually the same as the previous result. The maximum and minimum temperatures agreed to within 0.8 percent. The x stress using Mesh 2 is shown in

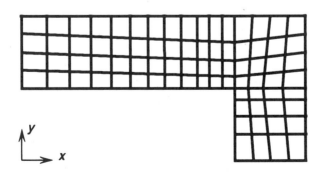

Figure 12-23. Finite Element Mesh 2

Figure 12-24. Now the stress concentration shows clearly although it is not very large. Also, the values along the inner and outer surface of the long leg have decreased, and the boundary condition for *x* stress on the right edge matches better. The Von Mises equivalent stress plot in Figure 12-25. looks more reasonable with an increase in the maximum value due to the inside corner stress concentration.

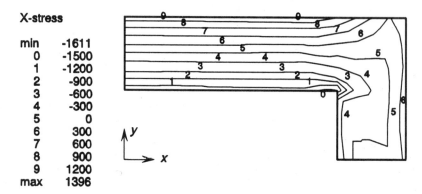

Figure 12-24. The X Stress Component in Mesh 2

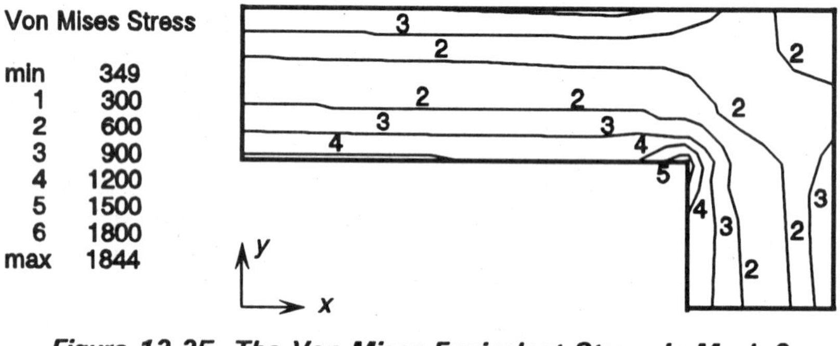

Figure 12-25. The Von Mises Equivalent Stress in Mesh 2

There is further refinement in Mesh 3 shown in Figure 12-26 again with the element size biased toward smaller elements at the inside corner.

The temperature results are within 0.2 percent of the previous results. The *x* stress plotted in Figure 12-27 shows a further increase in the stress concentration. Of course, the actual stress concentration is infinite for a perfectly square inside corner. From a practical standpoint there will be some finite radius or fillet in the corner and the results in these meshes represent a feature roughly equal to the element size. The boundary value of *x* stress on the right side now fits well. Stress values along the long leg

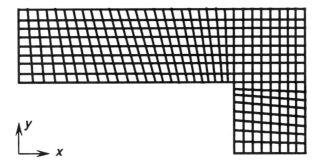

Figure 12-26. Finite Element Mesh 3

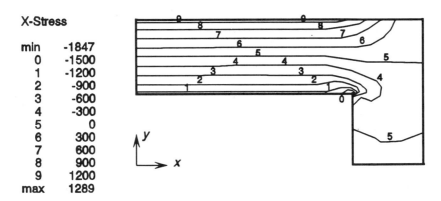

X-Stress

min	-1847
0	-1500
1	-1200
2	-900
3	-600
4	-300
5	0
6	300
7	600
8	900
9	1200
max	1289

Figure 12-27. The X Stress Component in Mesh 3

except for the inside corner did not change much so we have reached a reasonable convergence there. The Von Mises equivalent stress in Figure 12-28 shows a similar convergence except for the inside corner value.

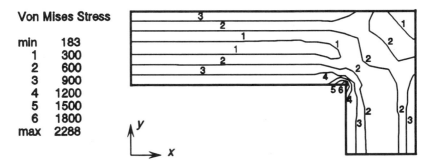

Von Mises Stress

min	183
1	300
2	600
3	900
4	1200
5	1500
6	1800
max	2288

Figure 12-28. The Von Mises Equivalent Stress in Mesh 3

Although general conclusions cannot be drawn from one case, this case shows that convergence of stress results required much more mesh refinement than did the temperature results. In most cases this will be true due to the nature of the solutions. The heat transfer solution finds the scalar values of temperature, while the stress solution first finds the vector components of displacement and computes the strain and resulting stress from derivatives of the displacement vector.

Problems

12.1 The conduit used in Problem 5.13 carries fluid with a temperature of 150°C. The extruded cross section shown in Figure P12-1 is A03560-T6 aluminum which has a thermal conductivity of 200 W/(m·°C). The fluid pressure is 4 Mpa. If the inside convection coefficient is 100 W/(m²·°C) and the outside coefficient is 5 W/(m²·°C) to air at 25°C, find the steady state temperature distribution and resulting thermal stresses. Also, determine the combined thermal and mechanical stresses and factor of safety against yielding. Compare these with the results from Problem 5.13 if it was done.

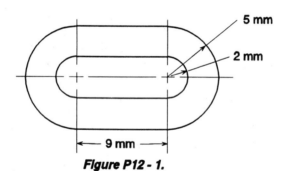

5 mm

2 mm

9 mm

Figure P12 - 1.

12.2 The injection molding cylinder flange from Problem 5.12 illustrated in Figure P12-2 carries a thermal as well as mechanical load. The bore temperature is constant at 500°F and the outside diameter value stays at 100°F. Thermal conductivity of the steel in the flange is 20 Btu/(hr·ft·°F). Determine the temperature distribution and thermal stress. Is the thermal stress significant compared with the mechanical stress due to internal pressure of 15 kpsi?

12.3 Recall the high pressure thick-wall cylinder design from Problem 7.3 in Figure P12-3. Using the dimensional data from Problem 7.3 determine the thermal stresses when the bore temperature is 600°F,

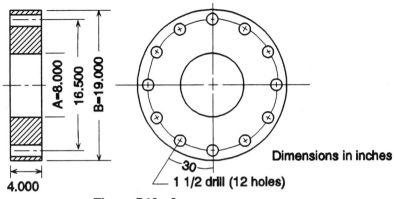

Figure P12 - 2.

the cooling groove surface temperature is steady at 50°F and the outer surface is insulated. Thermal conductivity of the steel in the vessel is 20 Btu/(hr·ft·°F). How do these stress levels compare with the comparable solid cylinder when the bore temperature is 600°F and the outer surface temperature is steady at 50°F?

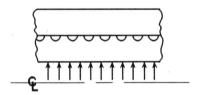

Figure P12 - 3.

12.4 The pressure vessel of Problem 7.5 is shown again in Figure P12-4. It contains a hot liquid that keeps the inside surface temperature at 650°F. An insulation layer that is 1 in. thick covers the outside surface and the bottom of the vessel. The convection coefficient from the outside of the insulation is 1 Btu/(hr·ft²·°F) to air at 70°F. Thermal conductivity of the steel in the vessel is 20 Btu/(hr·ft·°F) and of the insulation is 0.1 Btu/(hr·ft·°F). The stiffness of the insulation material is so low that it has no structural significance. Determine the temperature distribution and resulting thermal stress. Does this level of thermal stress have any significant effect on the structural performance of the vessel?

12.5 The thick hub rotor of Problem 7.8 is assembled with a long shaft as shown in Figure P12-5. The assembly calls for an interference fit of 0.003 in. Both the rotor and shaft are steel. Determine the interference fit stresses by thermally expanding the shaft elements

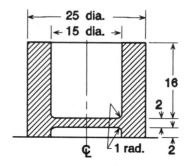

Figure P12 - 4.

or thermally shrinking the rotor elements. Find the rotational speed at which separation of the rotor and shaft would occur assuming the material did not yield.

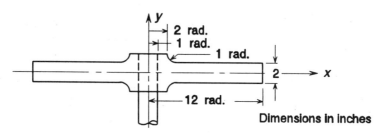

Figure P12 - 5.

References

12.1 Incropera, F. P. and DeWitt, D. P., *Fundamentals of Heat and Mass Transfer*, Third Edition, John Wiley and Sons, New York, 1990.

12.2 Boley, B. A. and Weiner, J. H., *Theory of Thermal Stresses*, John Wiley and Sons, New York, 1960.

12.3 Huebner, K. H. and Thornton, E. A., *The Finite Element Method for Engineers*, John Wiley and Sons, New York, 1982.

12.4 Cook, R. D., Malkus, D. S., and Plesha, M. E., *Concepts and Applications of Finite Element Analysis*, Third Edition, John Wiley and Sons, New York, 1989.

12.5 Wilson, E. L., and Ibrahimbegovic, A., "Use of Incompatible Displacement Modes for the Calculation of Element Stiffnesses or Stresses," J. Finite Elements in Analysis and Design, V. 7, No. 3, pp 229-241, 1990.

12.6 Roark, R. J., and Young, W. C., *Formulas for Stress and Strain*, Fifth Edition McGraw-Hill, New York, 1982.

APPENDIX

This user's guide is written to describe the operation of the FEPCIP, FEPC, and FEPCOP programs. It assumes the user has been instructed in use of the finite element method to solve stress analysis problems. The programs will be described with their capabilities and general sequence of use for solving a problem. The first time user should refer to the README file on the program disk for initial setup. In some parts of this guide the user is referred to more complete documentation in the file on the program disk.

These FEPC programs are provided as shareware for educational purposes. They are copyrighted programs and you are not authorized to sell or distribute them for COMMERCIAL purposes. You are free to use, copy and distribute them for NONCOMMERCIAL uses only if no fee is charged for use, copying or distribution. Specifically the programs were designed for use by students in university courses.

If you are an instructor and use the programs in a university course, I would appreciate a simple registration (no fee) of your university, course name or description, and approximate number of students taking the course.

If you use the programs for COMMERCIAL purposes, i.e., employed engineering, consulting, sponsored research, etc, a partial registration fee to continue the software development would be appreciated. For a full registration fee, you will receive the latest version, fully dimensioned, to run on an IBM PC or compatible with 640K RAM. Please state the current version number of the software you are presently using. Send inquiries to:

Dr. C. E. Knight
Professor of Mechanical Engineering
914 Ballard Ct.
Blacksburg, VA 24060

The programs are continually under development and your comments concerning present features or future enhancements would be appreciated.

USING FEPC, FEPCIP, AND FEPCOP

FEPC is a program that performs finite element stress analysis of two-dimensional truss, beam, plane solid, or axisymmetric solid structures. There are two companion programs. FEPCIP is the FEPC INPUT PROCESSOR which is used to input and check a model and prepare data files for FEPC. FEPCOP is the FEPC OUTPUT PROCESSOR which reads FEPC output data files and produces graphic displays.

The shareware version of the FEPC programs are currently dimensioned to run in a PC with 256K memory. The dimension limits in the programs are 250 nodes, 250 elements, 10 materials, 100 points, 30 lines, and 20 arcs. Also, the overall model size is limited in FEPC based on number of nodes and average nodal bandwidth. For example, 250 nodes with an average nodal bandwidth of 7, 200 with 10, or 150 with 14 are all maximum model sizes that can be run in FEPC. Dimension limits of the automatic mesh generation grid in FEPCIP are $I = 15$ by $J = 25$. Therefore a model can be built in FEPCIP even within the 250 node and element limit which is too large to run in FEPC, so plan carefully.

The procedure for solving a problem is to run FEPCIP to create the model, run FEPC to solve the equations, and run FEPCOP to display the results. The FEPCIP program presents the user with menus for interactive input, checking and storing a model. This creates an analysis file used as input for FEPC. Running FEPC produces a listing file of printout results and files of results used as input for FEPCOP. Graphic displays of deformed shape and stress plots may then be produced by FEPCOP.

ENTERING THE MODEL IN FEPCIP

Before starting to enter the model, develop a node and element numbering plan, boundary conditions, and the load placement for the model. With the FEPCIP.EXE file in the current drive and directory, begin by typing

FEPCIP < CR >

where < CR > means press the enter or return key. After the FEPCIP logo appears the program will continue after a short pause.

The screen will clear and the program will automatically detect the proper graphics mode for the supported graphics cards. The supported cards are IBM CGA, EGA, VGA, and MCGA of the PS/2 model 30 along with the Hercules graphics card.

After making the selection the main menu and graphics windows will then appear along with a prompt to SELECT A FUNCTION KEY.

INPUT PROCESSOR   FINITE ELEMENT PERSONAL COMPUTER   DATE TIME
TITLE:

F1 FILES F2 MODEL DATA F3 2D AUTOMSH F4 TITLE F6 CLEAR MEM F7 EXIT F8 VIEW OPTS F9 DSPLY OPTS SELECT A FUNCTION KEY	MODEL GRAPHICS WINDOW	MODEL SUMMARY WINDOW

Selection of a menu item by its function key brings up a branch menu for many of the selections.

Key F1 branches to a menu for recalling a previously stored model, storing a new model, or adding a title.

Key F2 branches to a menu for entering or editing all data required for the model.

Key F3 branches to a menu for two-dimensional area mesh generation.

Key F4 prompts the user to input a title for the current model.

Key F6 will clear all the current model data from memory in order to start entering a new model.

Key F7 exits the program.

Key F8 branches to a menu to change the current view of the model.

Key F9 branches to a menu to change the visibility of entities (nodes, elements, loads, etc.) or labels (node numbers, element numbers) on the next redraw of the model.

Every branch menu has a function key selection to return to the previous menu. Many of the selections on the branch menus will branch to additional menus. In each case, following completion of tasks on the current menu, use the previous menu selection to step back through the menus until the modeling is complete.

The general procedure for entering a model is to use the MODEL DATA function key to access the menu for selecting the element type, defining the material properties, defining nodes and elements, setting node displacement restraints, and applying loads. For truss and beam element models all the model data are entered from this menu and its branch menus.

Two-dimensional solid models using plane stress, plane strain, or axisymmetric elements may first use the 2D AUTOMSH selection to generate the model mesh of nodes and elements. Once the nodes and elements are defined return to the model data menu to complete the model by material definition, setting node restraints and loads.

The model must be stored on a disk file before exiting the program. The model data may be saved to disk at any time in the progress of building the model. Two files are stored for all complete models under a user specified filename with file extensions of .MOD AND .ANA. If the model is incomplete only the .MOD file is stored and messages denoting the yet to be defined data for the .ANA file are displayed. All the current model data is saved in the .MOD file.

The program operates by using the function keys to select the operation from the menu. When data is required a prompt appears on the data entry line just below the TITLE: header. The user types in the requested data separated by commas or spaces followed by the carriage return or enter key.

When the prompt to detect an entity appears on the data entry line a cursor will appear. If a mouse exists and its driver is loaded then use the mouse to position the cursor and press the left button to detect or the right button to abort. Otherwise with no mouse use the arrow keys to position the cursor at the entity location and press the space bar to detect or press the return or enter key to abort the detect and terminate the current operation.

FILES

```
F1 RCL FN.MOD
F2 STO FN.MOD
      & FN.ANA



F10 PREV MENU

SELECT A
FUNCTION KEY
```

Selecting FILES from the main menu branches to the submenu on the left. A previously formed and stored model may be recalled from disk by selecting F1 RCL FN.MOD. The user is prompted to enter the filename, FN, without its .MOD extension. The filename may include the drive designation and path, but it may be a maximum of 20 characters long including the drive designator characters. DO NOT enter any leading blank spaces in the input of the filename. If the file cannot be found an error message is displayed.

The current model may be stored on disk by selecting F2 STO FN.MOD & FN.ANA. The model currently in memory will be stored in FN.MOD assuming no errors. Also, if the model is complete and ready for analysis the input file for the FEPC program will be stored in FN.ANA. If the model is incomplete the user is given messages indicating which data are missing. The store operation may be done at any time during the progressive construction of the model in order to have a place to restart in case of destruction of the current model data in memory. If the files FN.MOD and FN.ANA already exist on the disk they may be overwritten with the user's consent by the current data in memory each time the store function is executed.

After completing use of this branch menu select F10 PREV MENU to return to the main menu.

MODEL DATA

Begin entering a new model by selecting F2 MODEL DATA on the FEPCIP main menu. This produces the branch menu shown below.

F1 ELEM TYPE
F2 MATL PROP
F3 NODE DEF
F4 ELEM DEF
F5 RESTRAINTS
F6 LOADS
F8 VIEW OPTS
F9 DSPLY OPTS
F10 PREV MENU
SELECT A
FUNCTION KEY

If a truss or beam element model is to be entered then all the model data will be entered from this menu. If a two-dimensional solid is to be entered then all the data may be entered from this menu or the 2D AUTOMSH selection may be used to generate the nodes and elements for the mesh. Once these are generated they may be edited from this menu. If 2D AUTOMSH is to be used it should be done first or after the element type is selected and material property sets are defined since it will overwrite any existing node and element definitions.

ELEMENT TYPE

The F1 ELEM TYPE selection displays the list of available elements. Use the indicated function key to select the element for the model. Only one element type may be used in a model. After selection the program returns to the previous menu. See the documentation file on the program disk for more detailed instructions.

MATERIAL PROPERTIES

The F2 MATL PROP selection branches to a menu for input and query of material set definition. Up to 10 material property sets may be defined and should be defined in numerical order. Material property sets may include some physical properties depending on the element type. Each element in the model has a material set number associated with it which defines its material properties. If different material or physical properties exist in different parts of the structure then multiple material sets should be defined before elements are defined so that the correct assignments may be made at the time of element definition. See the documentation file on the program disk for more detailed instructions.

NODE DEFINITION

The F3 NODE DEF selection branches to a menu to perform node operations. Nodes for truss and beam element models will all be defined in this section. If the 2D automesh option is used for the 2D plane and axisymmetric models then that should be done first and any additional node operations will be done in this section. Node operations include definition, generating a row of nodes between two defined nodes, moving, deleting and querying nodes.

A node is defined by its number and coordinate position. A prompt will appear on the data entry line to input a node. Simply enter the node number, and its X and Y coordinates. The data must be separated by commas or spaces. All leading blanks are ignored, but do not enter any trailing blanks. If the node is generated properly, it will be displayed on the graphics screen if it lies inside the current window. The starting window is 10 units by 10 units, but it will change automatically if any of the view options are exercised. Autoscale will resize the window so that all currently defined nodes fit inside. The prompt recycles so that the next node may be input. Terminate input by tapping the return or enter key. See the documentation file on the program disk for more detailed instructions.

ELEMENT DEFINITION

The F4 ELEM DEF selection branches to a menu to perform element operations. Elements for truss and beam element models will all be defined in this section. If the 2D automesh option is used for the 2D plane and axisymmetric models then that should be done first and any additional element operations will be done in this section. Element operations include selecting material, definition, generating a row of elements from a starting element, modifying, deleting, and querying elements.

Single elements are defined by user selection of nodes for each element. The user is prompted to detect each node needed for the element definition. The order of node selection on two node elements is of no consequence. *The nodes for four node elements must be picked in a counterclockwise order surrounding the element area.* Elements are numbered in numerical order as they are defined. Their material set assigned is the current material set. Each element is drawn on the graphics screen as it is defined. The node prompts recycle to define the next element and will continue until the right mouse button is pressed or the return or enter key is pressed at the node detect prompt to terminate element definition. See the documentation file on the program disk for more detailed instructions.

RESTRAINTS

The F5 RESTRAINTS selection is for applying node displacement boundary conditions. By default all nodes displacement components are free to take on nonzero values appropriate to the structure response under load. The components which must be zero for the model to behave properly are specified to be fixed. The menus that appear allow the user to set values for the restraints and then pick the nodes to which the set values apply. See the documentation file on the program disk for more detailed instructions.

LOADS

The F6 LOADS selection is for applying loads to the model. Loads may be node forces or element edge pressures (for 2-D solid elements). The menus that appear for node forces allow the user to set component values for the loads and then pick the nodes to which the set values apply. Edge pressure is applied by input of the pressure value and then selection of the elements to which the set pressure applies. Menu selections also allow deletion or query of forces and pressures. See the documentation file on the program disk for more detailed instructions.

2D AUTOMESH GENERATION

This section of the program is used for area mesh generation of two-dimensional plane stress, plane strain, or axisymmetric models. The principle of the approach is a mapping of an integer area grid into the geometrical area of the model. The geometrical area is defined using point locations, lines, and arcs. The perimeter of the geometrical area is defined by the complete set of lines and arcs which enclose the area.

The integer area grid will have lines which correspond to the lines and arcs of the geometrical area. Plan the correspondence by imagining or physically sketching on square grid paper the perimeter in the integer area. Use integer coordinates I and J with a range of 1 to IMAX and 1 to JMAX respectively. IMAX and JMAX values are listed in the first section of this guide. A 1 by 1 square in the integer area grid will map to an element in the geometrical area. Grid points in the integer area grid will map to node points in the geometrical area model.

Lines in the integer area can only be lines of constant I or Lines of constant J. The perimeter must be defined by a head-to-tail connection of lines in a counterclockwise(ccw) direction around the area. The length of line in the integer area is equal to the number of elements desired along the corresponding line or arc in the geometrical area.

The process involves defining the geometrical points needed to describe the model area, then defining lines or arcs using those points which complete the model perimeter. Next plan the corresponding integer area grid to be mapped into the geometry of the model.

After all the geometric points, lines, and arcs have been entered, area mesh generation can begin. The genmesh function presents a prompt to pick the starting point of the area. This point on the geometry will correspond to the 1,1 I,J coordinate location on the integer area. A series of prompts then proceed for the detection of a line or arc, the number of elements on that line or arc, and the direction of the corresponding line in the integer I,J area.

The first line or arc detected must have the selected starting point as one of its end points. The next line or arc picked must have the other end point of the first line or arc as one of its end points. Each successive line or arc picked must then connect to the other end point of the previous line or arc. This continues until the perimeter of the geometry is closed and the end point of the last line in the integer area must be back at the starting point, i.e., the perimeter in the geometry area and the perimeter in the integer area must close simultaneously. Both of these perimeters must progress ccw around the area.

When a line or arc is picked the entry of number of elements determines the length of the line in the integer area. The direction entry chooses one of four allowable line directions in the integer area. The directions are labeled 1, 2, 3, and 4, which correspond to right(+I), up(+J), left(-I), and down(-J) respectively in the I,J coordinates.

The integer area of the model must lie in the positive quadrant of I,J coordinates. Since the starting point in the integer area is at 1,1, and the perimeter must be ccw, the direction for the first line or arc must be 1. The direction for the second line or arc picked may be 1 or 2. Successive lines may have any direction values as long as some lines with directions 1 and 2 are used before any with directions 3 or 4 so that the I and J coordinate values always remain positive. The total number of elements on all lines in the 1 direction must match the total number in the 3 direction, and the total number in the 2 direction must match the number in the 4 direction.

The bandwidth of the structure stiffness matrix is minimized by making the number of elements in the I direction smaller than in the J direction. The limits are IMAX-1 elements in the I direction and JMAX-1 elements in the J direction. However, no model may have more than the maximum number of nodes or elements listed in the first section of this guide.

Mapping is an iterative process of distorting the integer area to fit in the geometry area. After a few iterations a mesh will be drawn on the screen. If it appears to be suitable then it can be accepted or more iterations may be requested to make it smoother. If it is unacceptable then a different integer area may be tried.

The menu of functions for mesh generation is

```
F1 POINT
F2 LINE
F3 ARC
F4 GENMESH


F8 VIEW OPTS
F9 DSPLY OPTS
F10 PREV MENU

SELECT A
FUNCTION KEY
```

POINT

Points are used to define lines and arcs which make up the model's geometric perimeter. Two points are needed to define a line, and three points along the arc are needed to define an arc. Points are input by their coordinate location. Selection of key F1 POINT brings up a submenu for creating, modifying or deleting points. See the documentation file on the program disk for more detailed instructions.

LINE

A straight line may be used to represent all or part of any straight edge on the model. More than one line on an edge might be used to produce different element spacings along the edge. If two or more lines are used on any single edge, they should be connected in series with no overlap. Selection of key F2 LINE produces a submenu for creating, modifying or deleting lines. A line is defined by picking two points at the ends of the line. See the documentation file on the program disk for more detailed instructions.

ARC

An arc may be used to represent all or part of any circular arc on the model of 180 degr. or less included angle. If more than one arc is used on a circular arc of the model then they should be connected in series. Three points along the arc are needed for the definition. They are the two end points and an intermediate point. Another point is created during definition of the arc at the arc's center of curvature. This may cause the autoscale function to reduce the model scale substantially if the arc radius is very large in order to fit all the points on the graphics screen. Selection of key F3 ARC produces a submenu for creating, modifying and deleting arcs. See the documentation file on the program disk for more detailed instructions.

GENERATE MESH

Selection of the F4 GENMESH function begins a series of prompts and inputs to define the meshing area. If the element type has not been selected the element menu will be presented for a choice. If more than one material set has been defined then the prompt to enter material set number will appear. Enter the set number which will be assigned to all the elements defined using the mesh generator.

Following these conditional entries the prompt to detect the start point appears. This is a geometric point on the model which corresponds to the 1,1 point in the I,J integer area. Next the prompt to detect a line or arc begins the sequence of perimeter definition. Detect the line or arc which connects to the starting point and starts on the ccw path around the perimeter by positioning the cursor on the line or arc center and pressing the spacebar or the right mouse button. Following detection enter the number of elements along the line or arc. Then enter the direction number of the line in the integer area(1, 2, 3, or 4). The set of prompts to detect line or arc, enter number of elements, and enter direction all cycle until the user terminates input by entry of the return key.

The user should be sure the geometry perimeter is closed before terminating. The program checks that the integer area is closed, and if so begins iterating on the mapping. This may take a few minutes. If the integer area is not closed a program message reports this condition and the geometry is redrawn. Another trial to input the perimeter may begin with selection of the automesh function.

If the integer area did close, after a few iterations the mesh is drawn on the screen with the prompt for more mesh iteration (Y or N). If it needs additional smoothing enter Y. More iterations will be done and the prompt will reappear. If it looks acceptable or it is to be redone differently then enter N.

The next question is it ok to keep (Y OR N). Enter Y to keep the mesh, or enter N to discard this mesh and redo the GENMESH function with another plan.

Once an acceptable mesh of nodes and elements is kept, return to the MODEL DATA menu to apply the displacement boundary conditions and loads, and perhaps define a material property set.

When the model data is complete go to the files menu and store the model and analysis files to disk. This should also be done periodically during building of the model in case of unexpected termination of the session or a different path is chosen to complete the model.

VIEW OPTIONS

This selection appears on many of the branch menus to allow exercising the view options without retracing the menus to reach them. The following menu appears on selection of view opts.

```
F1 AUTOSCALE
F2 ZOOM
F3 MAGNIFY
F4 CENTER


F10 PREV MENU

SELECT A
FUNCTION KEY
```

AUTOSCALE

Selection of autoscale automatically scales the graphics window to include all currently defined points and nodes.

ZOOM

This function allows the user to select a portion of the graphics window which is then scaled to fit the full window. The user is prompted to detect the two corners of the zoom area.

MAGNIFY

This option changes the size of the model displayed. The user is prompted to enter the magnification factor. A positive value must be entered; values larger than one will increase the size of the drawing and values smaller than one will decrease the size.

Subsequent use of the magnify command will enlarge (or decrease) the model display with respect to its current size. For example, magnifying your model by two and then by three produces an image six times larger than the original.

CENTER

The model may be moved by selecting a new center of the graphics window. The user is prompted to locate the new center

DISPLAY OPTIONS

Display options control which entities and labels are visible when a graphics plot is done. A branch menu appears.

```
F1 ENTITY SW
F2 LABEL SW

F4 MONO/COLOR


F10 PREV MENU

SELECT A
FUNCTION KEY
```

ENTITY SWITCH

An entity switch setting is off or on to control the individual entity's visibility. Selection of this function brings up a submenu listing all the entities and prompting the user to select an entity.

Selection of a function key will produce a prompt to change its current setting by default. A Y or return key entry will accept the prompt question.

LABEL SWITCH

This function controls the display of labels(numerals) for nodes and elements on the graphic model. A submenu allows selection of a function key producing a prompt to change its current setting by default. A Y or return key entry will accept the default.

MONO/COLOR

This function switches the display between black and white or color when the computer has a VGA graphics board. On other graphics boards the display is always black and white. Switching the VGA to black and white allows the screen graphics to be dumped to a black and white printer without loss of character intensity as sometimes happens when screen dumping color graphics to a black and white printer. Since only the drawing color palette is changed with this switch the change occurs when the next drawing is done after the switch. Execute the function key again to return to a color display.

THE ANALYSIS BY FEPC

When a model has been developed and saved, it is complete and ready to be processed by the finite element processor, FEPC. After exit from FEPCIP, and before starting FEPC, be sure that the filename.ANA file can be accessed by FEPC by copying it to the FEPC diskette in the same directory where FEPC.EXE resides, or by using the drive and path designation in the filename.

RUNNING FEPC

With the FEPC.EXE file in the current drive and directory, begin by typing

FEPC < CR >

After the FEPC logo appears, a prompt will appear to enter the model filename (with drive and/or path designation but without the .ANA extension),(20 characters max).

As the computations proceed, messages will appear on the screen reporting the computation step in progress. If errors occur, error messages will also appear on the screen. FEPC creates some other files as it runs. There is a listing file of all the printed output labeled filename.LST. This file should be studied by the user after an analysis to check the input data interpreted by FEPC and all the numerical output. A file labeled filename.MSH stores the node and element data for FEPCOP. A file labeled filename.NVL stores the node displacement and element stress data for FEPCOP.

Some other files are also created during the FEPC run which are deleted upon normal termination of the program so the disk which stores the .ANA file should have some excess space for these files during runtime. If the run terminates abnormally some of these files may still be on the disk with extensions of .ELM and .LOD. These and other output files will be overwritten when running a model with the same filename.

If the FEPC run was successful then FEPCOP may be used to display the results in graphic form. If the run was not successful then examine the filename.LST file for data errors or error messages that may help to correct the model.

The output from a FEPC run is stored in a listing file called filename.LST, where the filename is the same as the model file name entered when beginning the FEPC analysis. This file includes a listing of all the input data as well as all the numerical results. See the documentation file on the program disk for more detailed instructions.

FEPC ERROR MESSAGES

1 - OUT OF SPACE, MODEL IS TOO LARGE (I)
> The model is too large to run in FEPC. Reduce the model size in FEPCIP and try again. Consult the program limits given before.

2 - NODE 'n' HAS BEEN PLACED ON AN INCLINED BOUNDARY, BUT IT IS ALREADY CONSTRAINED AGAINST X OR Y DISP. OR BOTH
> An inclined boundary angle is specified for node n, but an x or y restraint was also specified which is incompatible. Edit the model in FEPCIP.

3 - FEPC.EXE file not found
> The FEPC.EXE file must reside in the current drive and directory to execute.

4 - 5 No longer used.

6 - 'm' ELEM IS HIGHER THAN NO. OF ELEMENTS IN THE GROUP
> The filename.ANA file has been corrupted because element number m is higher than the total number of elements. The *.ANA file is an ASCII file so it may be printed or edited. Examine its contents in comparison with the data printout in the filename.LST file.

7 - ELEMENT NO 1 IS NOT DEFINED FIRST
> The filename.ANA file has been corrupted because element number 1 is not defined first in the list of element definitions. The *.ANA file is an ASCII file so it may be printed or edited. Examine its contents in comparison with the data printout in the filename.LST file.

8 - YOU HAVE A ZERO LENGTH ELEMENT #'m'
> A beam or truss element is defined using the same node for both ends or the two nodes defining the element have coincident coordinate locations.

9 - BAD ELEMENT #'m'
> A quadrilateral element is improperly defined or is too distorted. Check for cw node order definition around the element(it should be ccw), inside angles between sides greater than 180 degrees, butterfly shaped element, or a triangle formed by using one node for two corners (this is legal if the last two nodes in the element definition are the same).

10 - STIFFNESS MATRIX NOT POSITIVE DEFINITE, NEGATIVE STIFFNESS DIAGONAL TERM FOR EQUATION 'n', VALUE = '#'
> During solution of the system equations a negative diagonal term is found which means that the equations cannot be solved. The equation number corresponds to the free node degree-of-freedom in the system ordered consecutively with node numbers. These are listed in the filename.LST file produced in the FEPC run. Find the node number from this list then examine the elements which are defined using this node number for errors. If the equation number is 1 or the last equation number then the error is probably due to lack of sufficient displacement restraints to prevent rigid body motion.

11 - INCLINED BOUNDARY ANGLE MUST BE BETWEEN -89.99 AND +89.99
DEGR
  The inclined boundary angle input is outside the allowable range.

GRAPHIC RESULTS USING FEPCOP

After a successful run by FEPC, results files, filename.MSH and file-
name.NVL, will have been created on the disk. These are the input files
for output processing by FEPCOP.
  With the FEPCOP.EXE file in the current drive and directory, begin by
typing

  FEPCOP<CR>

where <CR> means press the enter or return key. After the FEPCOP
logo appears the program continues after a short pause.
  The screen will clear and a prompt will appear to

ENTER MODEL FILE NAME (NO EXT) -

Enter the filename (with drive and/or path designation but without the
.MSH or .NVL extension),(20 characters max). Some messages will appear
noting the progress of calculations, and then the screen will clear and the
FEPCOP MAIN MENU is displayed along with a prompt to SELECT A
FUNCTION KEY.

```
F1 DEFORMED
F2 X-STRESS
F3 Y-STRESS
F4 XY-STRESS
F5 T-STRESS
F6 VON MISES
F7 TRUSS STRS
F8 BEAM STRS
F9 OPTIONS
F10 EXIT


SELECT A
FUNCTION KEY
```

Selection of F1 DEFORMED brings up a branch menu.
Selection of F1 PLOT in this branch produces a deformed
shape plot of the element mesh superimposed over the
undeformed model. This plot shows the finite element
mesh when the node displacements are scaled and added
to the node coordinates so that the deformed shape is
exaggerated. In truss and beam models the deformed
mesh is superimposed over the undeformed mesh plot.
In 2-D solid models the deformed mesh is superimposed
over the outer boundary of the undeformed shape. The
displacement scale factor may be changed in the OP-
TIONS menu to increase or decrease the plotted deformation. An
additional submenu appears for view options of the plot.
  Selection of F2 ANIMATE in the branch menu produces a sequential
mesh plot of truss and beam models or a boundary outline plot of 2-D
models showing the progressive deformation as the load is cyclicly
applied. Press any key to terminate the animation.

The next five function key selections on the main menu show the stress contour plots developed in 2-D solid models for the indicated components of stress. Each function is accompanied by a legend of the contour values and the view options menu. The T stress component is nonzero only for the axisymmetric element models and represents the hoop stress in the axisymmetric structure. The Von Mises equivalent stress is calculated based on the distortion energy failure theorem using all the stress components calculated in the loaded model.

TRUSS STRS

This function is used to display the results in truss element models. The user may select a plot of the axial force or stress in all truss elements. The plot is in a bar chart format with the heights scaled to the maximum value in any element. Plus or minus signs are drawn on the bar near the top to indicate whether the member is in tension or compression.

BEAM STRS

This function is used to display the results in beam element models. The user may select to plot the axial, flexure, average transverse shear, or the maximum combined axial plus flexure stress. These are also in bar chart format with signs indicated near the top of each bar. The axial and transverse shear stresses are constant along an element length so one bar per element is sufficient. The flexure stress component varies linearly along the element length so a bar is plotted for the value at each end. Two bars are also plotted for the combined axial plus flexure stress. The sign of the combined stress is the same as the sign of the axial stress which is the combination producing the largest magnitude.

OPTIONS

```
F1 NODE SW
F2 ELEM SW
F3 DISP SCALE

F10 PREV MENU

SELECT A
FUNCTION KEY
```

Function key F9 OPTIONS produces a submenu. Selecting F1 adds node symbols to the 2-D stress plots, and F2 adds element outlines inside the 2-D boundary for stress plots. Both of these selections produce a prompt to switch the current setting. Answering Y or by default a return key entry will make the change. Selection F3 allows the scale factor for the deformed shape plots to be changed by prompting for a new scale factor with the current scale factor shown as the default value. Enter a larger value to increase the exaggeration or a smaller value to decrease it.

INDEX